77th Conference on Glass Problems

77th Conference on Glass Problems

A Collection of Papers Presented at the 77th Conference on Glass Problems, Greater Columbus Convention Center, Columbus, Ohio, November 7–9, 2016

Edited by
S. K. Sundaram

The American Ceramic Society

WILEY

Published by John Wiley & Sons, Inc., Hoboken, New Jersey.

Published simultaneously in Canada.

Library of Congress Cataloging-in-Publication Data is available.

ISBN: 978-1-119-41779-8
SSN: 0196-6219

10 9 8 7 6 5 4 3 2 1

Contents

Foreword

The 77th Glass Problem Conference (GPC) is organized by the Kazuo Inamori School of Engineering, The New York State College of Ceramics, Alfred University, Alfred, NY 14802 and The Glass Manufacturing Industry Council (GMIC), Westerville, OH 43082. The Program Director was S. K. Sundaram, Inamori Professor of Materials Science and Engineering, Kazuo Inamori School of Engineering, The New York State College of Ceramics, Alfred University, Alfred, NY 14802. The Conference Director was Robert Weisenburger Lipetz, Executive Director, Glass Manufacturing Industry Council (GMIC), Westerville, OH 43082. Donna Banks of the GMIC coordinated the events and provided support. The themes and chairs of five half-day sessions were as follows:

Operations
Elmer Sperry, Libbey Glass, Toledo, OH and Uyi Iyoha, Praxair Inc., Tonawanda, NY

Controls
Glenn Neff, Glass Service USA, Inc., Stuart, FL and Bruno Purnode, Owens Corning Composite Solutions, Granville, OH

Energy
Phil Tucker, Johns Manville, Denver, CO and Martin Goller, Corning Inc., Corning, NY

Melting
Jans Schep, Owens-Illinois, Inc., Perrysburg, OH and Andrew Zamurs, Rio Tinto Minerals, Greenwood, CO

Forming
Kenneth Bratton, Emhart Glass Research Inc., Windsor, CT and James Uhlik, Toledo Engineering Company, Inc., Toledo, OH

Refractories
Laura Lowe, North American Refractory Company, Pittsburgh, PA and Larry Mc-Closkey, Anchor Acquisition, LLC, Lancaster, OH

Preface

This volume is a collection of papers presented at the 77th year of the Glass Problems Conference (GPC) in 2016. This conference continues the tradition of publishing the papers that dates back to 1934. The manuscripts included in this volume are reproduced as furnished by the presenting authors, but were reviewed prior to the presentation and submission by the respective session chairs. These chairs are also the members of the GPC Advisory Board. I appreciate all the assistance and support by the Board members.

As the Program Director of the GPC, I am thankful to all the presenters at the 77th GPC and the authors of these papers. This year's meeting concludes with a record total attendance of 497 and also 44 students. I appreciate all the support from the members of Advisory Board. Their volunteering sprit, generosity, professionalism, and commitment were critical to the high quality technical program at this Conference. I also appreciate continuing support and leadership from the Conference Director, Mr. Robert Weisenburger Lipetz, Executive Director of GMIC and excellent support from Ms. Donna Banks of GMIC in organizing the GPC. I look forward to continuing our work with the entire team in the future.

Please note that The American Ceramic Society and myself did minor editing and formatting of these papers. Neither Alfred University nor GMIC is responsible for the statements and opinions expressed in this volume.

S. K. Sundaram
Alfred, NY
November 2016

Acknowledgments

It is a great pleasure to acknowledge the dedicated service, advice, and team spirit of the members of the Glass Problems Conference (GPC) Advisory Board (AB) in planning this Conference, inviting key speakers, reviewing technical presentations, chairing technical sessions, and reviewing manuscripts for this publication:

Kenneth Bratton—Emhart Glass Research Inc. Hartford, CT
Warren Curtis—PPG Industries, Inc., Pittsburgh, PA
Martin Goller—Corning Incorporated, Corning, NY
Uyi Iyoha—Praxair Inc.,Tonawanda, NY
Michelle Korwin-Edson**—Owens Corning Composite Solutions, Granville, OH
Robert Lipetz—Glass Manufacturing Industry Council, Westerville, OH
Laura Lowe—North American Refractory Company, Pittsburgh, PA
Larry McCloskey—Anchor Acquisition, LLC, Lancaster, OH
Glenn Neff—Glass Service USA, Inc., Stuart, FL
Adam Polcyn*—PPG Industries, Inc., Pittsburgh, PA
Bruno Purnode—Owens Corning Composite Solutions, Granville, OH
Jans Schep—Owens-Illinois, Inc., Perrysburg, OH
Elmer Sperry—Libbey Glass, Toledo, OH
Phillip Tucker—Johns Manville, Denver, CO
James Uhlik—Toledo Engineering Co., Inc., Toledo, OH
Justin Wang*—Guardian Industries Corporation, Geneva, NY
Andrew Zamurs—Rio Tinto Minerals, Greenwood, CO

In addition, I am indebted to Donna Banks, GMIC for her patience, support, and attention to detail in making this conference a success.

* Joined the AB at the 76th GPC.
** Joined the AB at the 77th GPC.

Operations

GLASS PLANT AUDITS – THREE CASE STUDIES IN GLASS PRODUCTION PROBLEMS AND THEIR SOLUTIONS

J. M. Uhlik
Toledo Engineering Company, Inc.
Toledo, OH

ABSTRACT

Over the multi-year length of a glass plant campaign, problems arise with root causes traceable to design, engineering, construction and the operational parameters which can change over time. This presentation's intent is to illustrate real-world production problems arising from those changing needs, practical solutions, and the value of non-resident process reviews such as performed by the member companies of the TECO Group. It often takes an experienced or non-routine study of the problem(s) to first determine the root cause, and then engineer how to best resolve it. Problems and solutions can include: refractory design/selection for maintenance, wear issues experienced during the campaign, hot repairs and temporary engineering solutions and operational process adjustments. This can typically result in glass quality improvements and campaign life extension through applying principles of operation optimization and improving maintenance techniques. The results are often significant improvements in glass quality, pack yields and the plant's bottom line. This presentation will discuss three problem instances - in a throated furnace, the waist area of a float furnace, and sidewall refractory replacement maintenance activities.

INTRODUCTION

Ask anyone involved in the day to day operation of making glass - sometimes it seems as if their plant is a living, breathing entity. And sometimes, they become ill...

GLASS PROBLEM ONE – FURNACE WITH A SORE THROAT

TECO was asked to assist with an increasingly evident non-conforming glass attribute from a throated furnace in Europe. The problem was a distortion line in the rolled glass sheet being called a "water mark" by plant personnel, which tested as a high density alumina-zirconia layer approximately 60-75 microns thick, shown below in Figure 1.

When first detected preliminary thinking was that it was a lamination problem (mechanical action on the glass), such as roller mark, lip issue, roller cooling problems, etc. Many initial actions were undertaken to find the root cause and eliminate this defect. These actions included:

- The cover of the lamination area was adjusted.
- Various machine positions were instituted.
- Several machine changes with different rollers were tested, smaller rollers with different cooling, etc.
- A bottom roller with chrome coating was used.
- Refractory lip was changed.

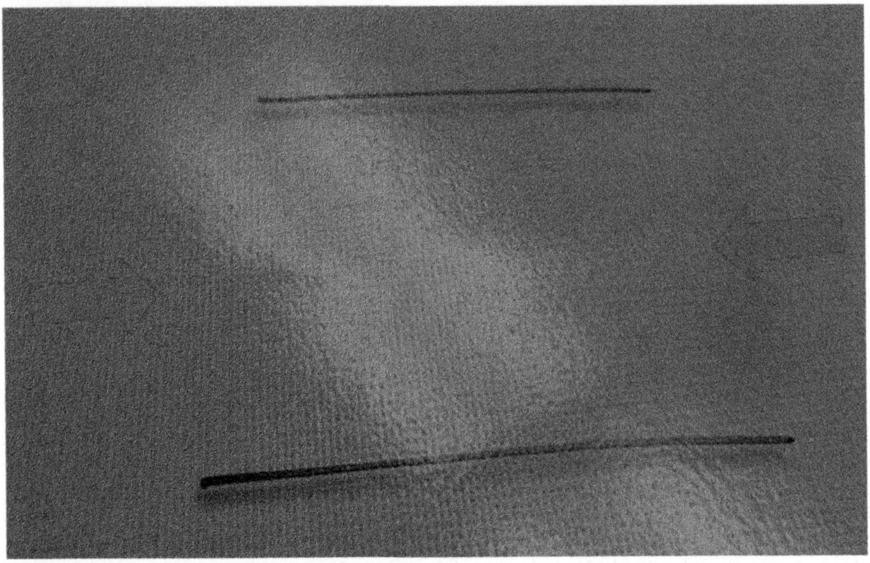

Figure 1. Optical inhomogeneity in the Ribbon

This is a typical operational progression, where the urgency of continuing glass losses force increasingly costly (in terms of lost production and/or equipment replacement) adjustments to the process in a search for improvement. Meanwhile a sample of the distortion line was sent out for laboratory analysis. The results are shown in Figure 2.

Based on the analysis report, an average of five composition measurements yielded higher levels of alumina and zircon content than what was normally found in the base glass. Therefore, increased focus was placed on the batch, the glass furnace and the forehearth operation and structures, which had been previously been operated consistently and at steady state for some period of time.

Technical service personnel from Toledo Engineering Co., Inc. (TECO) and Zedtec, Ltd. were invited to the facility to help the customer assess the situation. Together, the combined team completed several problem solving exercises and developed an evaluation plan. During this investigation, the physical inspection of the furnace interior was performed, as the viewing ports allowed. Figure 3 shows the interior of the Zedtec glass conditioning forehearth - the inspection of the forehearth provided assurance that there was no undo wear, the structure was intact and the glass level was as per the design of 50 mm below top of block.

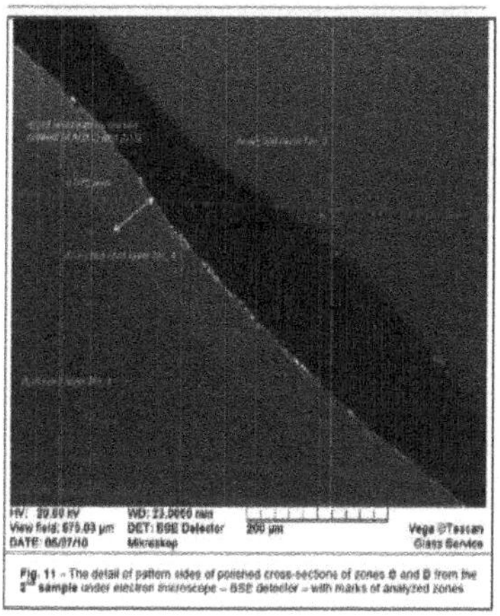

Formula	Ox%
Na_2O	14,76
Al_2O_3	3,37
SiO_2	74,10
Ca0	7,04
ZrO_2	0,74

Figure 2. Analysis Results of the Glass Inhomogeneity.

Figure 3. Forehearth Inspection Port and Forehearth Glass Level Estimate

Finally, the inspection of the furnace interior provided that while the structure and superstructure refractory appeared to be in proper condition, the glass level as observed did not appear to be at the design level of 50 mm below top of block – there appeared to be much less glass freeboard, as shown in Figure 4. To check this observation, first a simple length of tubing was used as a water level, and when checked, showed that the furnace construction was correct, with both the furnace and forehearth top of block set to the same elevation. The actual glass level observation did not make sense, so not only was the water level used several more times, but an optical engineering level measurement was contracted locally, and these readings also verified the correct construction. Engineering 101 teaches us that liquids seek their own level, yet the visual observations appeared contrary to this. The team assembled and discussed the next steps.

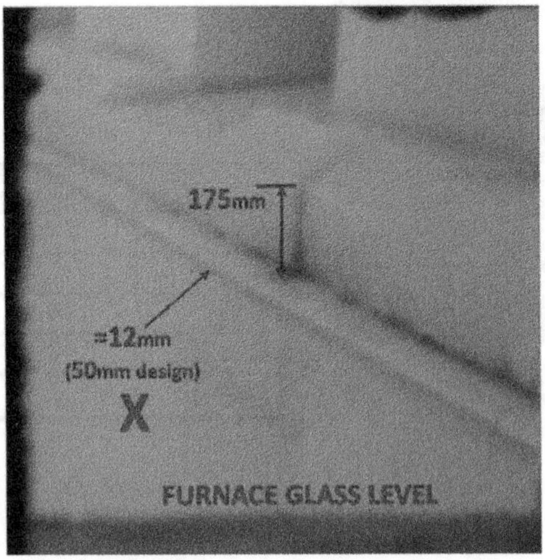

Figure 4. Furnace Glass Level Visual Estimate

Although seemingly improbable, a theory developed that perhaps there was restriction in the throat, possibly a buildup of denser glass that was 'wicking off' and presenting in the final product as the aforementioned watermark. The throat became the focus of the discussion, and a plan was developed to retrofit a drain onto the throat bottom, to remove a possible accumulated buildup of denser glass:

Plant management acted quickly to institute this solution. The results after draining the throat for a few hours, during which periods of inhomogeneous glass streaming were evident, was that the furnace glass level returned to the designed 50 mm below top of block. While seemingly improbable, an accumulation of denser glass in the throat area had slightly restricted the glass flow, requiring a higher furnace glass level and head pressure to maintain the operating glass level in the forehearth.

The distortion line in the glass was the presentation of this problem - a buildup of denser glass which restricted glass flow - and was solved by installation of a periodic drain capability in the bottom of the furnace throat, as shown in Figure 5.

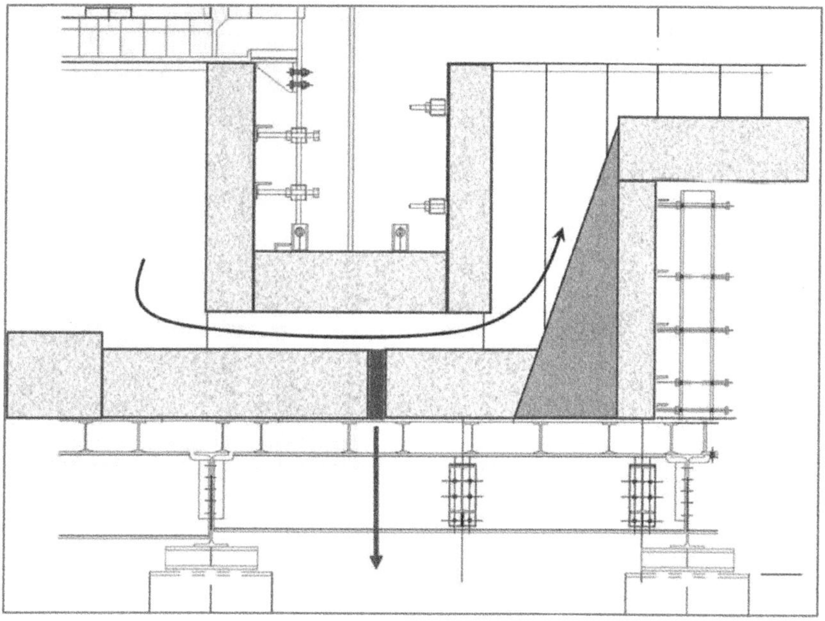

Figure 5. Representation of a Sunken Throat Bottom Drain, such as by KTG Engineering.

GLASS PROBLEM TWO – FURNACE WITH A SAGGING WAISTLINE

TECO was asked to assist a float glass manufacturer who had recently changed a large refractory structure in the waist area of their float furnace, to relieve a possible source of refractory contamination in their glass ribbon. In normal circumstances, this should be a straightforward procedure, the replacement of the A arch (see Figures 6 & 7 below).

The A arch, as can be seen in the Fig. 1, is a high and narrow design that helps shield the downstream area of the waist during normal openings of the upstream access area, in front of A, for routine maintenance in that area. The old A arch, replaced by the customer, is shown in Figure 8.

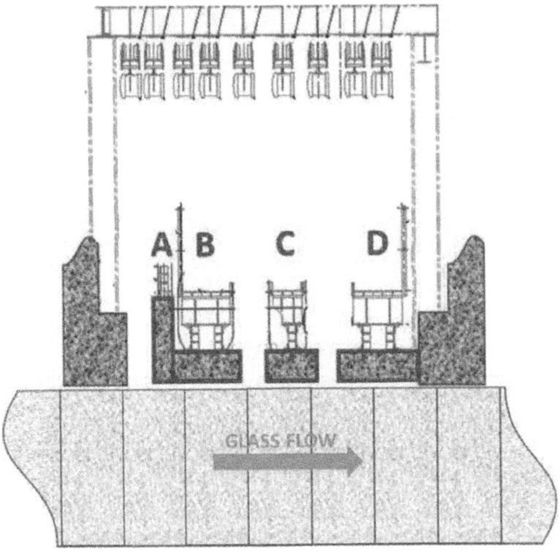

Figure 6. Layout of Waist Arches A through D

Figure 7. View of Waist, Right Side

Figure 8. Old Replaced A Arch

However, during the replacement of the A arch, the support structure of the B arch was exposed to higher temperatures and radiant heat from the open A arch area. This is normally acceptable for the short period of the A arch replacement procedure, in that the B waist arch support steel is designed to be water cooled. Unfortunately, the steel assembly provided by a local supplier had developed water leaks when originally placed in service, and the B arch support beam was necessarily switched over to compressed air cooling to avoid leaking water damage to the refractory structure. Periodic inspections of the B arch had shown only slight sagging (Figures 9 and 10) while being cooled with compressed air, and it had remained stable for several years.

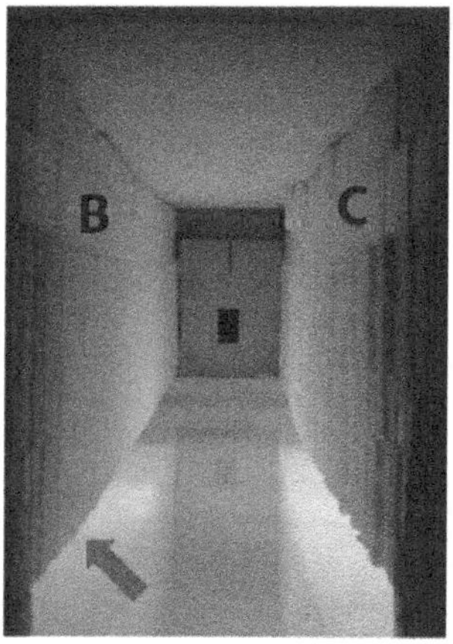

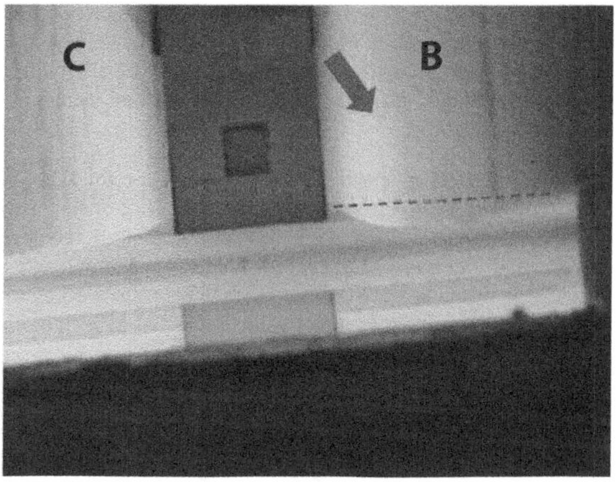

Figure 9. View from Right Side

Figure 10. View from Left Side

When the A arch was replaced, the B arch support was exposed, became overheated, and sagged severely during the replacement work, as shown in Figure 11. The B arch became a possible risk to the safe and efficient operation of the float glass process line going forward.

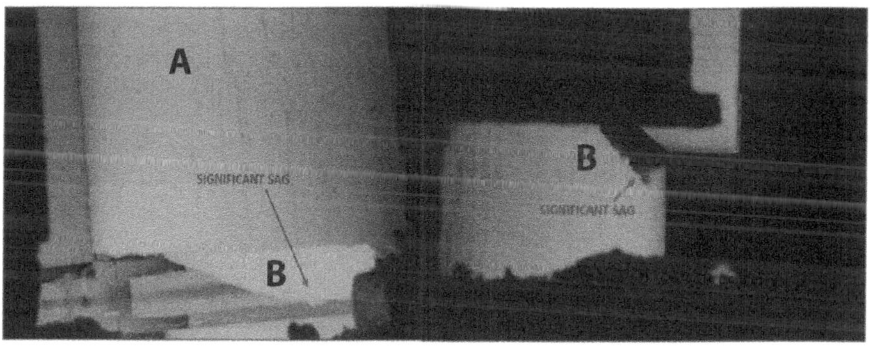

Figure 11. Views of the B Arch Maximum Sag during Replacement of the A Arch

With the discovery of the damaged B arch, TECO was asked to provide its expertise and participate in the emergency plan for the replacement of the B arch, which also supports an equipment access walkway above the arch. In general, this waist area is a fairly crowded space (see Figure 7 above). The new B arch refractory assembly had to be carefully preheated in order to sustain its introduction into the elevated temperature of its position in the furnace waist area. The procedure which was developed by the team was to transfer the new, preheated B arch into position simultaneously with the removal of the old B arch, and with the reintroduction of water cooling of the new B arch steel support structure. Therefore, the team carefully considered all aspects of personnel safety and staffing, mechanical structures, refractory heating requirements, piping, equipment logistics, and operational adjustments, along with risk assessment and contingency planning.

The new B arch was preheated in a temporary kiln built proximate on the operating floor, and when fully heated it was insulated, lifted out of the heating area, chained to the old B arch, and as the old arch was lifted and removed to the left, the new B arch followed from the right side into position. In Figure 12 below, the old B arch is shown when removed from the waist. The over-heated area of the beam is evident in the center, as well as some of the permanent sagging of the refractory that resulted. Figure 13 shows the new B arch in place.

Figure 12. Old B Arch Removal from the Waist, with New B Arch being Set in Place from the Right Side

Figure 13. New B Arch in place, with Water Cooling

The procedure was executed quickly and efficiently as planned, resulting in a new, secure and stable refractory structure in the float furnace waist area. No further issues were reported afterwards.

GLASS PROBLEM THREE – FURNACE HAVING ITS BOWLS TAKEN OFF

TECO was asked to assist a float glass manufacturer with the removal of the glass level bowls in their furnace design, when it was determined that these aging structures were highly worn and a source of glass loss. The physical removal work was to be performed during a routine maintenance hot-hold of the furnace. Working as a team, a concept was jointly developed and planned in detail.

Glass level bowls (GLB) have been built into furnaces in the past, primarily for the utilization of nuclear level detectors, and even farther in the past, as a structure called a dry dock alcove when ceramic floaters were used in the waist as a skimmer, prior to the adoption and use of water cooled waist pipes, as shown in Figures 14 and 15. As can be seen in Figure 16 below, the GLBs do not contain the full depth of the glass, however, their removal and replacement with standard height furnace sidewall refractory, meant that the entire depth of the molten glass in the furnace would be exposed, and therefore needed to be reliably constrained.

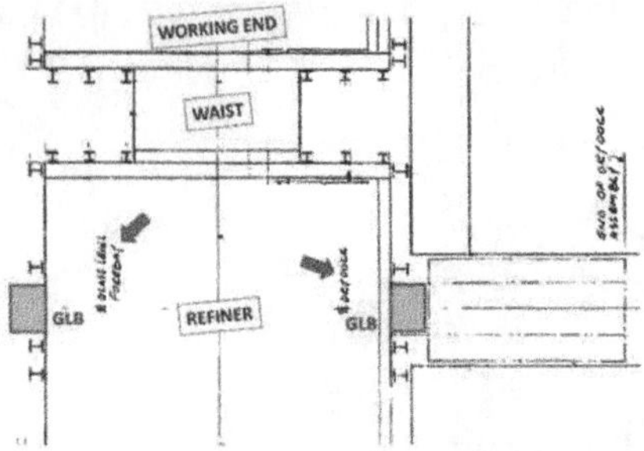

Figure 14. Typical Position of GLB Shown in an Old Furnace Drawing

Figure 15. An exterior photograph of the Old Furnace

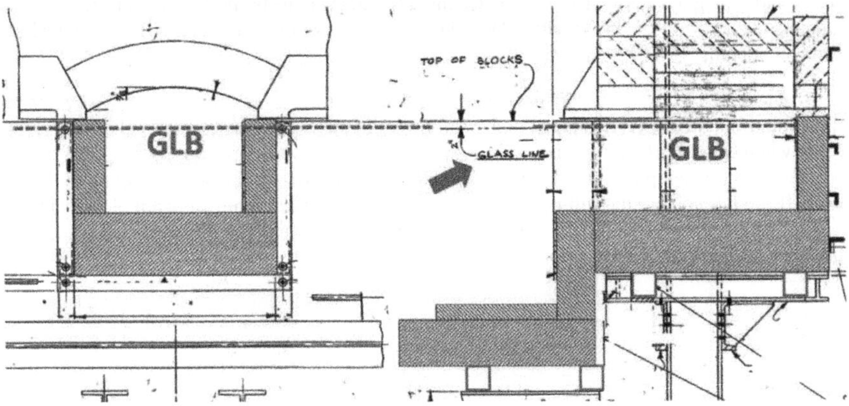

Figure 16. End View and Centerline View of a Typical GLB Construction

The glass level in the furnace was planned to be reduced only a few inches for the work, resulting in several feet of molten glass still requiring safe containment, as shown in Figure 17.

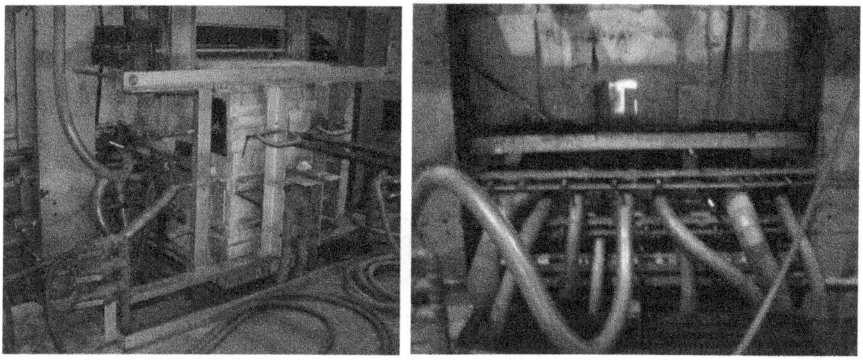

Figure 17. Left: Left Side GLB before the Work; Right: Work Nearly Finished after its Removal

To accomplish removal, a procedure to insert water lances directly into the molten glass was developed, in order to freeze the glass and secure the four to five-feet wide opening that was required. Water lances are used for similar applications in smaller furnaces, but this is atypical for a large float furnace containing approximately 1500 tons of molten glass. The application of the water lances also had to accommodate a clear working area for maximized personnel safety, and of course, facility protection from any molten glass leak. The steps of the work are highlighted in Figure 18.

Figure 18. Left: First Water Lance Readied after GLB Superstructure Removed; Right: inserted (R)

As a result of water being introduced directly into the molten glass depth, the steam being produced initially caused significant eruptions and displacement of the normally calm glass surface. The area of glass in the furnace in front of the GLB was monitored and evaluated for the planned for amount of cooling required to stabilize the area in order to contain the furnace molten glass and allow complete removal of the GLB and lower furnace sidewall blocks in this area. Once judged secure, the demolition was completed. These steps are presented in Figures 19 – 21.

Once demolition was completed, the furnace sidewall area was cleaned and prepared to accept the new replacement refiner tank wall blocks in this area. Half height blocks were utilized, in order to facilitate their safe and secure placement. Older furnace designs often used two course construction in the refiner areas, and this was judged acceptable for meeting the remaining campaign requirement.

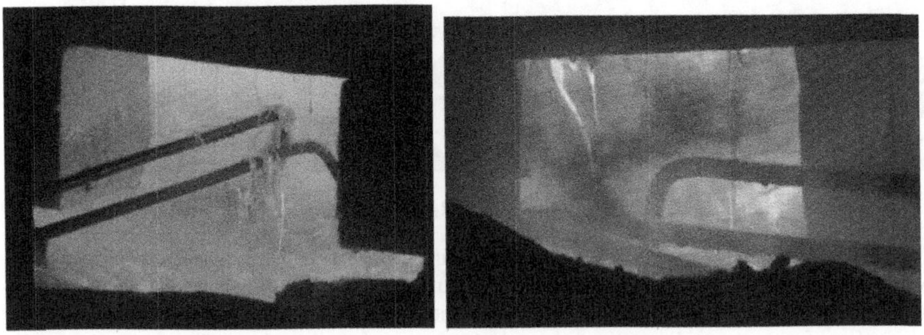

Figure 19. Second (L) and Third Lances (R) When Initially Placed

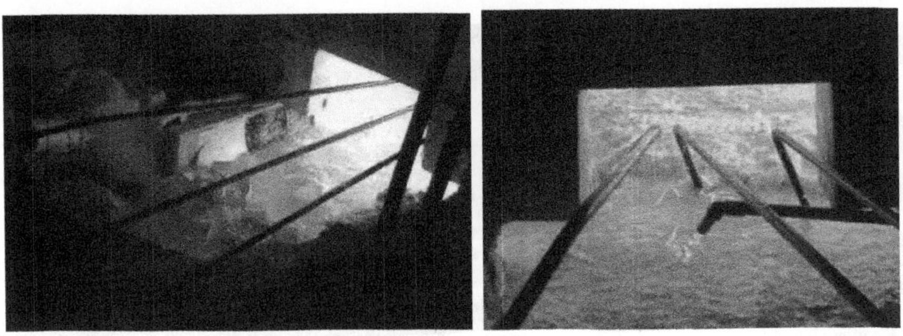

Figure 20. Water Lances Placed via. the GLB Opening

Figure 21. Left: Area of Concern in the GLB Furnace Inside Corner; Right: During Demolition. Bright orange spots of hot glass can still be seen on the right side.

After all refractory replacement work was completed and the steelwork secured, the water lances were removed and the same procedure was successfully followed once again to remove the GLB on the other side of the furnace. Figures 22-24 show the progress.

Figure 22. Left: First Block Placement; Right: Ready for the Last Sidewall Block

Figure 23. Interior View as Work Neared Completion. Note: Due to the rapid exposure of the large tank wall block into the operating furnace high temperature environment, some block corner spalls occurred.

Figure 24. Secure Replacement Refractory (shown at cold repair inspection, after three full years of service)

SUMMARY

These three examples are representative of situations encountered during a glass plant's operating life. They are somewhat 'episodic' and rare at each facility, in that these types of problems occur infrequently during a long furnace campaign. Glass plant operators may wish to team with qualified glass plant engineering companies, experienced in current 'best practices', for added expertise and guidance when needing to assess and solve such problems.

ACKNOWLEDGEMENT

TECO wishes to gratefully thank their customers, who kindly granted their permission to use this information.

FURNACE DESIGN AND OPERATION FOR THE LONG TERM

D. Boothe
Allstates Refractory Contractors, LLC
Waterville, OH 43566

ABSTRACT

Glass manufacturing companies are pushing the limits on gross tons of high quality glass they are demanding from their glass furnaces, via pushing the furnaces longer and harder. The expectations for more robust refractories, timely and economical hot repairs and stable quality are now a "given"! The return on investment is expected to surpass the espoused life of the furnace. In order to accomplish this demand for longer campaign life, numerous proactive models must be placed in motion, beginning before the start of the latest campaign. Selection of robust refractories, design and operating parameter computer control systems must be implemented well before the first glass bottle or lite of glass ever comes off the line. In situ monitoring of the key furnace parameters throughout the campaign is vital and the consequence selection of "on-the-fly" hot repairs and the capable contractor who is selected to perform these repairs are all extremely important variables that affect both furnace glass quality and campaign life.

INTRODUCTION

For the past 2 to 3 decades the glass industry has concentrated on better refractories, fuel economy, environmental issues, glass productivity and the ROI of any improvements. However, last on that list has been the consideration of the factors that all these efforts have on the effects on long-term operation and inevitable need for an interim repair typically while the unit is online.

GOOD OPERATIONS / PRACTICES

Let's deal with the effects of operations first. Many experienced operators believe that the greatest factor leading to longer furnace campaigns is in protecting the stability of operation. Upsets from tonnage changes, power failures, and equipment breakdowns can take a toll on the furnace life. Handling upset conditions influences approximately 1/3 of furnace life.

It has long been my opinion that glass furnace operations require a human presence and that human presence needs to be actively involved in the visual conditions inside the furnace, refiner or working end, and forehearths, carrying out handheld optical measurement of the various sections of these units and be the final decision maker on the operational changes. The industry has gradually abandoned this mode of operation to be more cost effective. This has led to an increase in catastrophic events; I believe lost production, glass quality, glass breaches and the lack of trained individuals to effectively combat these issues.

OPERATIONAL CONCERNS

Temperatures more than those required for adequate melting increase the "safety factor" regarding glass quality, but do not show up as a detriment to furnace life for many years.

Effective energy management in the control of a furnace has a significant influence on operating economics and consistent glass quality, but can also have serious detrimental effects on refractory and furnace life. Figure 1 shows energy consumption of glass melting furnces. In addition, flame control (length, velocity, redox and location) can have significant effects on refractory, carry over and thereby furnace life.

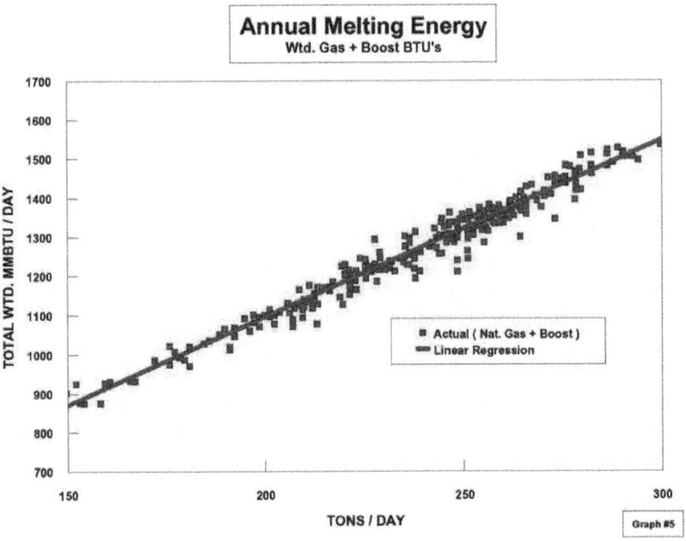

Figure 1. Energy Consumption vs. Tonnage

Each furnace and glass type has a quantifiable "energy characterization curve" which defines operating energy versus pull rates and cullet levels. Once defined in operation after a furnace rebuild, hourly or daily comparisons can be made between expected and actual fuel inputs, temperatures and glass quality to determine actions to be taken with energy adjustments.

One of the most difficult aspects to appreciate about a furnace operation is Thermal Momentum, as shown in Figure 2. The total mass of molten glass and hot refractories contribute a significant amount of energy to the melting process, but must be replenished by heat from flame combustion and electric boosting sources. When tonnage changes occur, a new equilibrium condition must be established. For even modest pull rate changes (5 to 10 %), this process can occur over several days. Observations regarding actual vs. expected energy input and glass bath temperature stability during a transition period are better understood by viewing the previous operating conditions over a period of days. It is for this reason that instrumentation using data storage and graphic presentation techniques help present the furnace operator with a better understanding of the melter's condition.

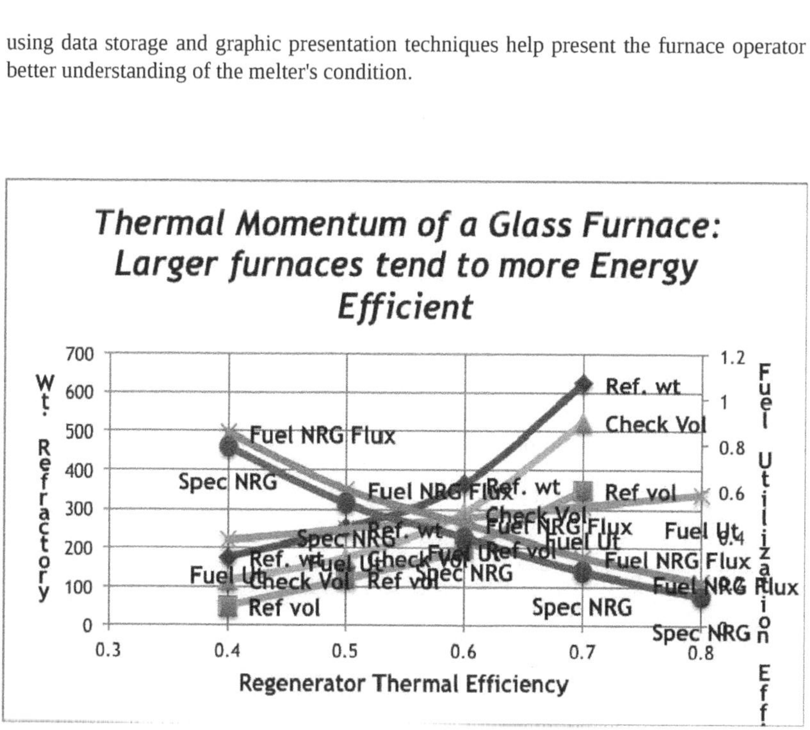

Figure 2. Thermal Momentum of a Glass Furnace

Changes in the batch pattern over a period of hours can indicate a changing thermal condition requiring operating adjustments. Higher pull rates naturally result in a larger percentage of the melter's area being occupied with batch. Similarly, changes in the cullet content can change the batch melting rate and influence the observed batch pattern. This can be significant, if the cullet is not mixed with the batch, but is added separately in the charging operation.

With the advent of internal cameras, infrared and improved thermal measurement, glass level, and combustion sensors, and computerized controls, the focus has been to minimize the human element in operations, but this, in my opinion, is a huge error in judgment. Cameras double as optical pyrometers and collect temperatures at specific spots. However, cameras are great for historical data on a generalized picture of the furnace and its operation, but are limited in scope of coverage and if not monitored constantly, can lead to catastrophic consequences. Sensors are snap shots of a single confined area; they deteriorate over time and require continual calibration to provide consistent accurate data. Computerized controls cannot make accurate decisions on the previously mentioned potential sensor issues and are not visually attuned to the conditions inside the furnace regarding batch coverage, raft melt out location, bubbler flow continuity, flame geometry and other influences that are humanly recognizable.

Glass manufacturing is a science, but it is also an art and we are systematically eliminating the art with increasingly poor results. My favorite question to corporate management is," If I pulled the plug on the electrical connection to all this instrumentation, how or better yet who would be around and know how to operate the furnace during this outage"? I was once told the 6 P's of success, Prior Planning Prevents Potentially Poor Performance! A common saying from the "Old Guys", "What's EXPECTED, gets INSPECTED". Things such as:

Checklists of procedures

Procedures based on Best Practice with an emphasis on why we do things, not just what to do.

Independent Audits to assure best practices are included and followed.

Or

Where is or how to start a backup generator, if one is available or does not start on its own and the frequency of testing online switching and operation.

Knowing where and how to affect a natural draft for the unit?

How to get heat into the furnace and forehearths without electricity?

Shutting down the glass flow, so as not to empty the forehearths.

Removing the charger or insulating it from the heat once the doghouse has melted out.

Review of how to handle the energy input without batch or at 100% cullet.

How to affect sidewall-cooling event the cooling fans are out?

How or where to switch from recirculated water to city water and periodic testing of switching.

Or

Having a very detailed shutdown procedure to save the furnace in a long-term power outage and how much annual detailed and hands on training is given to all crews?

These are not some hypothetical scenarios; they have happened to me personally several times in my 45+-year career.

TIPS AND TRICKS

Now let us move onto some other operational tips and tricks that hopefully will benefit those who are looking for exceptional furnace life, better glass quality and fewer surprises. To help predict refractory wear, potential glass leaks and major repair schedules, the collection of historical infrared temperature profiles is a positive step. This means rather than waiting until we begin to see the redness in refractory, we proactively scan the infrared profile of the entire furnace on a regular scheduled basis beginning with an initial scan at about 6 months into the furnace operations. This initial scan will provide a base line picture of the new furnace and its profile brand new. Figure 3 shows the need for periodic inspection.

This data collection should take place once each year for the first 5 years of furnace life. Then a scan every six months for the next three years. After that, dependent on the results of these scans, additional scans should be taken at 90 to 120 day intervals of the critical or excessive wear sections of the furnace.

In Depth inspections of the furnace structure are a necessity

To prevent disruptive breaches and downtime.

Figure 3. Left: Depth Inspection; Right: Disruptive Breach of Furnace

This should help schedule overcoats, preventative maintenance, minor or major repairs and most desirable prevent glass leaks. In the same vein, as previous comments about instrumentation, regular calibration of handheld optical temperature measurement devices is critical, as are oxygen and combustibles analyzers. Figure 4 shows representative temperature profiles. Every Critical Parameter should have redundant measurement or correlation factors to others.

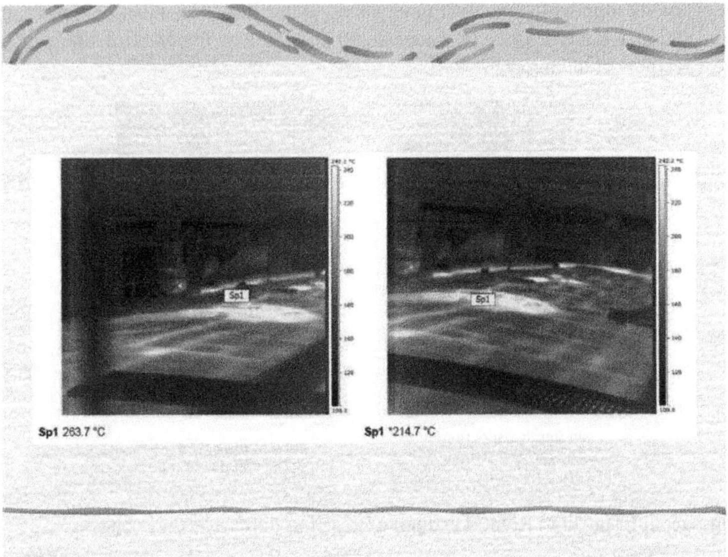

Figure 4. Thermal Profiles in a Furnace

An alternative to the infrared scanning of sidewalls, is an old trick of taking a hacksaw blade and inserting it into the joints of the sidewall (assuming they are exposed) and push it in until it stops (Figure 5). Then measure the length of the insertion vs. the known block thickness and record it and the location. This is a time-consuming process, but will give one an estimate of the glass penetration over the height of the block

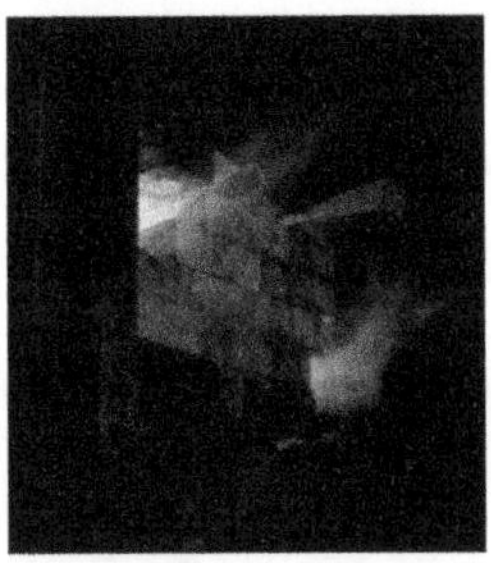

Figure 5. Thickness of Glass Penetration Measurement

Raw material quality, batch preparation, stored cullet and their transport to batch house and furnace are often overlooked and go unaccounted for. The quality of materials sent to the furnace is critical items and need inspection on a consistent basis. Regular batch samples should be taken at the charger to note possible segregation, cullet to batch raw material ratio and dispersal, moisture content and particle size distribution tests for all raw materials and visits to supplier facilities to confirm specifications and processing are necessary functions. Cleaning of conveyors, elevators and transition points are essential for good operation and can affect furnace operations and life. These are all indications of potential glass quality issues. Once again, proactive analysis can prevent glass upsets and surprise glass quality events. Additionally, glass oxide chemistry plotted by oxide and viscosity cures calculated to maintain melting, liquidus and forming temperatures.

For those of us who must answer to management when the furnace that was predicted to last 8, 10 or more years, didn't! The industry needs to convert furnace life expectations from years to tons pulled. Yes, it will upset the accountants and they must find a new formula for depreciation, but pulled tonnage for any specific glass color and quality is the single best predictor of furnace life. We know that the current average lifetime total tonnage pulled in a typical container glass furnace is approximately 1,200 to 1,500 tons per square foot of melt area per campaign. This figure represents a 25% increase over the average tonnage Fay Tooley reported 30 years ago.

For the float folks, the 1990's tons pulled for a float furnace was approximately 2,000,000 tons prior to rebuild and/or rebuild planning. Today's goals exceed 3,000,000 tons, as shown in Figure 6. It is so very important to have a baseline for the specific glass and furnace design. It is important here to note, very low iron glass compositions will skew these numbers and shortens furnace life even more.

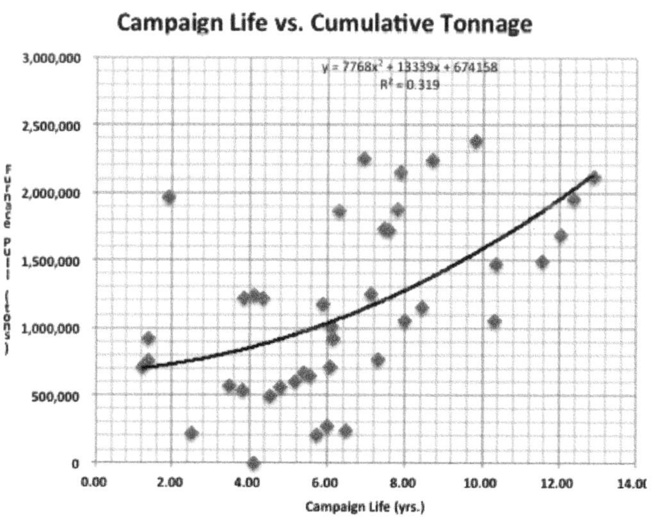

Figure 6. Cumulative Tonnage vs. Campaign Life

A habit that became a doctrine at Thatcher Glass where I began my career was that of 'Doc' Arrendale. Doc Arrendale was the senior scientist and VP of Thatcher Glass in the 60's, 70's and early 80's. That doctrine was the daily historical data collection of every furnace in the Thatcher organization. For decades, this was done by hand, graphed and cataloged. What this provided was a starting point for every job change on every furnace. These records were kept and organized per best productivity for any job run on each furnace during its campaign life. It gave temperatures, tonnage input vs. tonnage pulled, glass batch melt pattern referenced to distance from the charger down the furnace, forehearth and gob temperatures with the corresponding pack percentage. A daunting task by hand, but an ultimately valuable tool to the furnace crew to set up for a job change and subsequent best operational reference points. In today's world of computers and unlimited data collection and analytics, this task would be much simpler, quicker and easily available. Asahi Glass in Japan is currently practicing this type of data collection for its entire company and every operating furnace.

Preparedness for any eventuality is essential for the survival of any furnace. Having at least 2 water lances that will reach any portion of the furnace in the event of a leak and leak preparedness training of all furnace personnel and the local fire department is a necessity. This is covered in great detail in a paper given by Dan Carman at Hotbels 2016.

One more item that should be on the top of every furnace supervisor and maintenance crew's agenda is to replace a defective thermocouple or temperature measurement devices as soon as it ceases reading or appears erratic. The absence of a temperature measurement device, especially in a critical area such as the throat, crown or exhaust flue can be a monumental loss and lead to catastrophic results!

FURNACE DESIGN AND MAINTENANCE

The second part of this presentation is directed towards the engineering and design aspect of furnace life. As we go into the concept and design of a new furnace or rebuild, our normal approach is to study it's past performance, environmental footprint and future expectations. Utilizing tools like modeling, both mathematical and physical, vendor input for combustion, control and similar unit performance, refractory testing, etc., we attempt to provide a total production unit capable of handling the sought-after tonnage and quality predicted over its life cycle.

But do we consider the eventual need for midterm repair or overcoat? When GM developed the Northstar engine for the Cadillac, they had to insure it would run for at least 100,000 miles between tune ups, because to access the rear 4 sparkplugs, one must either remove the firewall in the passenger cabin or disconnect the engine from the transmission and frame mounts and roll it toward the radiator. This cost is extremely expensive, therefore the requirement for the long cycle between tune-ups. Figure 7 shows a design that helps accessing the furnace for repairs.

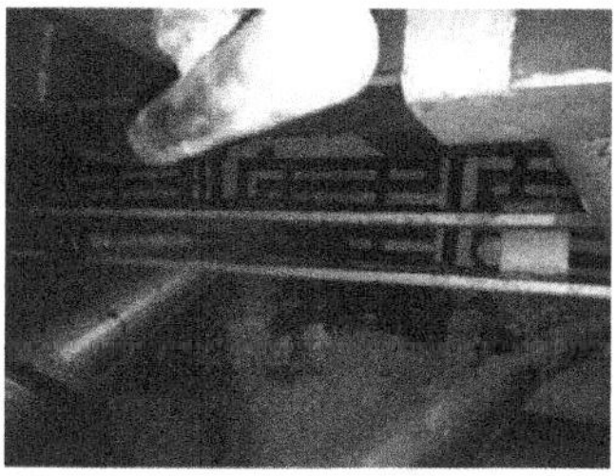

Figure 7. "Design" in Accessibility for Overcoat Repairs

A similar situation exists in our furnaces when we design the sidewall backup, insulation, binding and cooling. Very little thought is given to the well-known eventuality that and / or possibly several overcoats will be required. We do not consider or envision placement of binding steel to accept an overcoat with little need for steel revisions or access. Backup refractory and insulation sizes that will accept placement of typical size overcoat refractory within the existing joint structure; binding and minimal cutting are not generally a consideration. Cooling ductwork that has shutoff slide gates conveniently placed with slip joints to make removal and reinstallation easy and fast. Quick connects on water piping to remove water panels come under this heading as well.

Once a plan is in place for an overcoat then one should proceed from one end to the other. It is very difficult to overcoat in a location-by-location or skip fashion and then expect the best fill in or connecting installation with different crews over many months, seasons or years.

The use of all fused cast refractory in many of our furnaces especially oxy/fuel can make sidewall overcoats, throat overcoat or complete changes or any under glass line repair more than challenging. To access the inside of the furnace to successfully freeze the glass, water lances need to be introduced into the glass to prevent glass leaks during removal of back up materials and installation of the overcoat. This access is generally limited in an all fused cast furnace and the only alternative is to drill through the crown and insert the lances straight down the sidewall or in front of the throat or doghouse. This is not an easy task in an all fused cast furnace and not generally well accepted by most of our corporate engineering groups. Alternative access has not been a consideration in most of our furnace designs. However, lancing through the crown should be the last resort with some glass chemistries for it can create defects for many days or even weeks after completion. This is just another reason for preemptive wear studies and overcoat or cooling to reduce the need for freezing the glass. A few penny's worth of prevention is worth thousands in curing the problems incurred by waiting.

The addition of stubbed cooling duct locations for end of campaign hot spot cooling would be advantageous. The electrical cost of adding more cooling air late in the campaign is justified for the extension of refractory life at the metal line. Additional water supply and drain connections along the length of the furnace to add water-cooling panels to stem sidewall wear towards the end of the campaign are a good investment. For those who do not believe in water-cooling panels, they are the best alternative in critical areas that are difficult to reach during a campaign or after an overcoat in these areas to extend the life of those repairs. Additionally, water quality for these panels needs to be a major consideration! Lime build up or dirty water can affect the cooling capacity of the panels over time if the water quality is poor.

Timely refractory maintenance can contain some minor concerns and prevent them from becoming major ones, as shown in Figure 8. Unfortunately, these practices are usually lowest on the plant's and management's priority list.

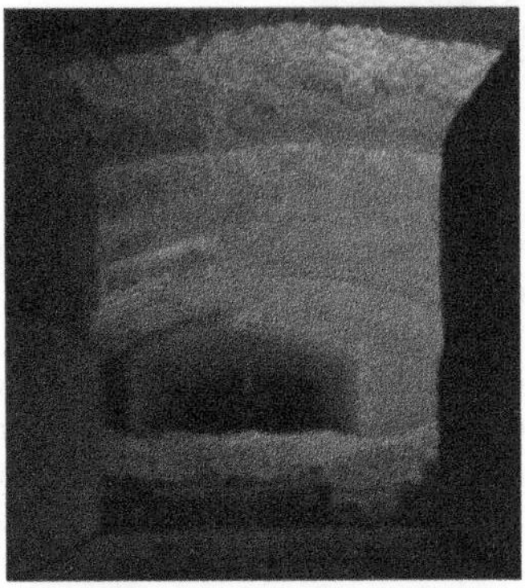

Periodic inspections and maintenance of checker pack and flues are a MUST.

Flue slag build up and checker plugging reduce furnace life and impact glass quality.

Figure 8. Left: Need for Periodic Inspections and Maintenance; Right: Flue Slag Build Up and Checker Plugging

Some of these additional designed in items will add cost to our capital projects. However, when one looks at the cost of hot repairs in later campaign years, many of these items will lower those costs substantially. Others will not add significant cost, but with consideration, again will lower the cost of later repairs and operations.

PLANNING FOR FURNACE REPAIRS

A few more comments regarding choosing a hot repair group. Hot repair and its method are very serious components of a furnace campaign. Have a detailed scope of work prepared for the project with all the information from the above-mentioned testing available for review. Knowing where the area may be weak or thin is a responsibility of the owner. Being prepared for the possibility of exposure to molten glass or potential collapse is imperative from both a personnel safety standpoint and the possibility of a catastrophic glass leak. It is so much more important to monitor the furnace and plan the work and work the plan. Figure 9 shows some examples of "on-the-fly" repairs.

Figure 9. Some Examples of "On-the-Fly" Repairs

To get the best outcome, it is essential that the company who is chosen for the project has a plan and experience for the specific area that is being repaired. Do not be averse to asking for a method statement not only from a safety standpoint, but also from an approach and finished product perspective. Do not assume that they have significant experience in this type of work. Review the plan, ask questions, and be comfortable with it before you release them to affect the repair.

CONCLUSIONS

I can see the wheels turning in you minds and the questions of "How do we do that when there may be glass running on the floor or refractory hanging precariously in or out of the furnace"? My purpose here is to try to prevent those issues from becoming the emergencies we have all seen too many of these past few years. Being proactive is far better and cost effective than reactive!

Controls

ENERGY REDUCTION WITH MODEL BASED PREDICTIVE CONTROL

M. Powys and D. Armagost
Gallo Glass Company, Inc.
Modesto, CA

M. P. H. Muijsenberg, R. Bodi, J. Muller, J. David, and G. Neff
Glass Service USA, Inc.
Stuart, FL

ABSTRACT

The paper will describe how Model Based Predictive Control (MBPC) was implemented at Gallo Glass in the USA on three (3) of their four (4) furnaces. Furthermore, MBPC is being implemented one of the new furnaces of Gallo Glass.

INTRODUCTION

A well designed furnace (and forehearth) can be controlled in the economical optimum with MBPC. MBPC optimizes energy input; keeping the balance between temperature stability, glass quality and energy consumption. Operating the furnace using flexible gas firing and electric heating/boosting with fully automatic MBPC allows the glass producer to operate the furnace in the most optimal cost-effective way. The technology can bring energy costs savings in the range of +/- three percent (3%).

ENERGY REDUCTION WITH MODEL BASED PREDICTIVE CONTROL

This paper will discuss the implementation of a Model Based Predictive Control (MBPC) system at Gallo Glass Company, Inc. (Gallo Glass). The MBPC system that was installed at Gallo Glass was the Glass Service, Inc. (GS) Expert System III™ (***ESIII™***) [referred herein as the MBPC Supervisory Control System].

Traditionally, standard PID control loops are used for fast processes with single input/output variables for almost the entire glass production line. However, glass melting is a very slow process (greater than 24 hours), which shows strong correlations between multiple input and output (MIMO) variables, including disturbance variables.

The MBPC system has the following features:

- The MBPC Supervisory Control System (MBPC System) is installed on top of the existing control station.
- The MBPC System can make automatic correlations that are needed for the control between all inputs and all outputs.
- The MBPC System provides for 24/7 consistent control and stabilization of the furnace operation, and especially the temperatures (for improved glass quality).

- An optimal balance between top gas firing and electric boosting can also be provided by the MBPC System, taking into account the actual natural gas and electricity prices to minimize energy costs.
- The MBPC System can also minimize emissions (i.e., NO_x and CO_2).
- The MBPC System can promptly react to sudden process changes (i.e., pull rate, cullet ratio, calorific heat value, etc.).

Gallo Glass is presently utilizing three (3) MBPC Systems on it's furnaces, with plans for a fourth furnace. More and more, MBPC Systems are being installed worldwide on the melters as an advantage for furnace stability and energy savings. Furthermore, and step by step, the MBPC Systems are also being applied to forehearths, but the basic functionality of the forehearth including gob weight control must be available. MBPC System control of a furnace is much like the autopilot features on airplanes today, and the future of self-driving cars that should be standard on most cars by the year 2025.

Figure 1, below, shows the main energy input parameters for an end-fired furnace. The MBPC System utilizes a matrix of dynamic response models that are generated automatically from the historical production database.

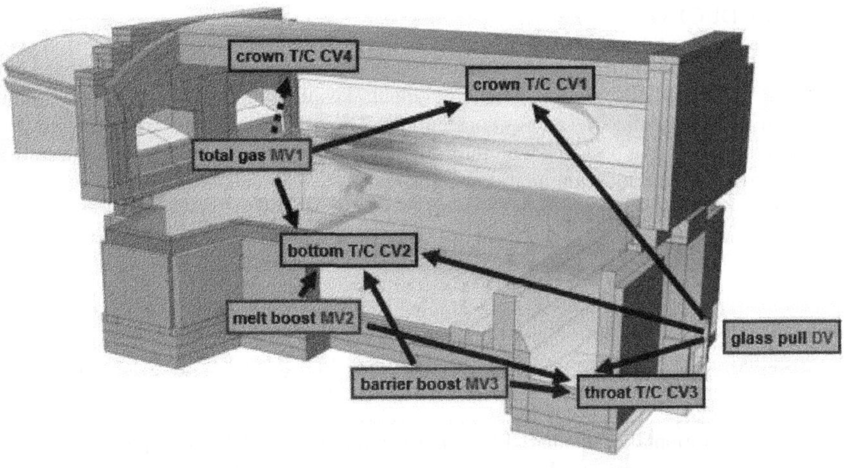

Figure 1. A MBPC System Operation Using the Energy Sources on a Furnace

Figure 2, below, shows how the models are developed. In this example the throat temperature increases with a reduced pull rate, while the melter bottom decreases. At the same time, the furnace melter crown temperature increases. The melter bottom temperatures are affected as well. The MBPC system will develop the models that will control the furnace based on these correlations.

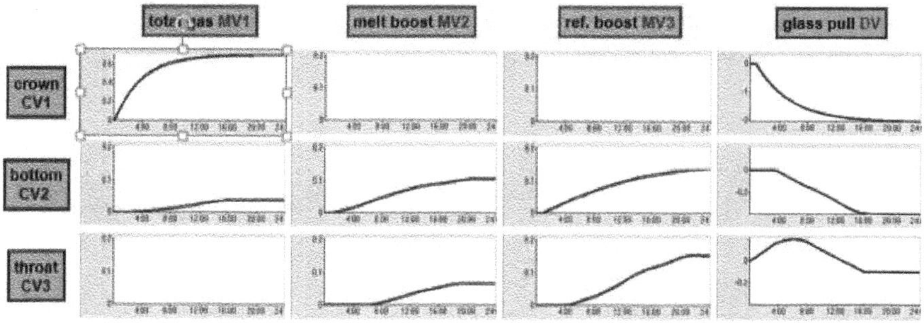

Figure 2 – A MBPC System Operation Using the Energy Sources on a Furnace

Figure 3, below, is a histogram of temperature improvements that can be made with the MBPC System for a glass furnace. In this case temperature variability has been reduced. This reduction in variability leads to the potential for energy savings if the operator reduces the set point.

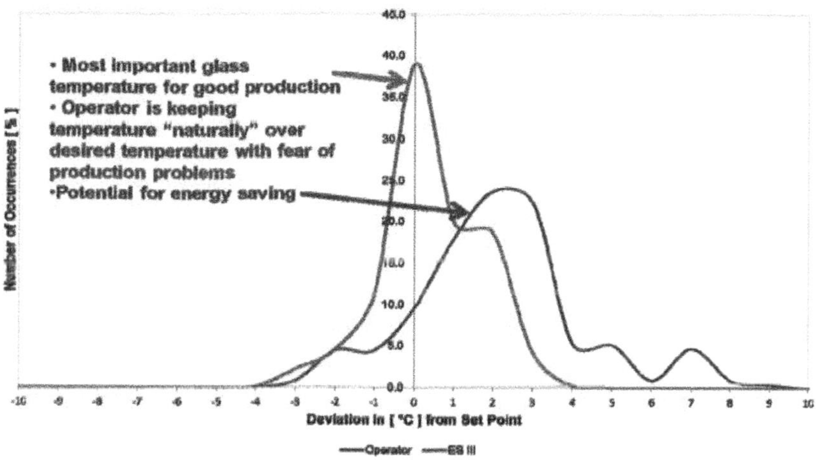

Figure 3 – Bottom Temperature Control Improved using MBPC

Figure 4, below, outlines the growth and development of E&J Gallo Winery, and in particular, Gallo Glass, Inc. The company was founded in 1933 followed by a period of growth for Gallo Glass from 1958 to 1982, where over the course of this time, five (5) furnaces were constructed. During the years from 1990 until 2000, the company pursued environmentally friendly changes, with furnace conversions to oxygen firing, with an electrostatic percipitator and scrubber added, as well as the use of more cullet in the glass melting process. During the years from 2000 until 2012, the Plant's capability was increased with outside glass sales, bulk transportation of the glassware, lightweighting of the glass, and the achievement of industrial safety protocols and certifications. Then from 2014 until present, the capacity of the Plant was improved with updated furnace designs, a new batch plant, triple gob IS (Independent Section) machines, and a warehouse expansion.

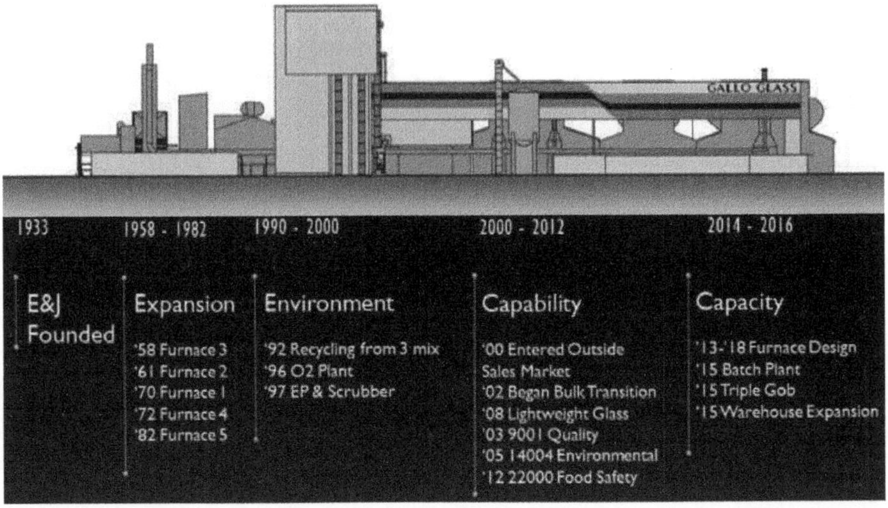

Figure 4. E&J Gallo Winery and Gallo Glass Plant History

We specialize in Wine and Spirits, with a premium image in light weight bottles for a product range from 187 milliliters up to 4 liters capacity. The Plant has installed triple gob wine bottle capability that will support the growth and investment in the wine and spirits segment to produce 100 million cases by the year 2023. To meet this demand, upgrades such as packaging flexibility and an in-house mold shop will be undertaken. These production increases will be mindful of environmental concerns, included BACT (Best Available Control Technology) such as oxygen fired furnaces, and exceeding the California EPA AB32 Regulation concerning CO_2 emissions, and the California minimum cullet recycling goal.

Figure 5, below, shows a few of the major wine and spirit bottles at the projected production rates, including flint, green, deadleaf green, and antique green bottles.

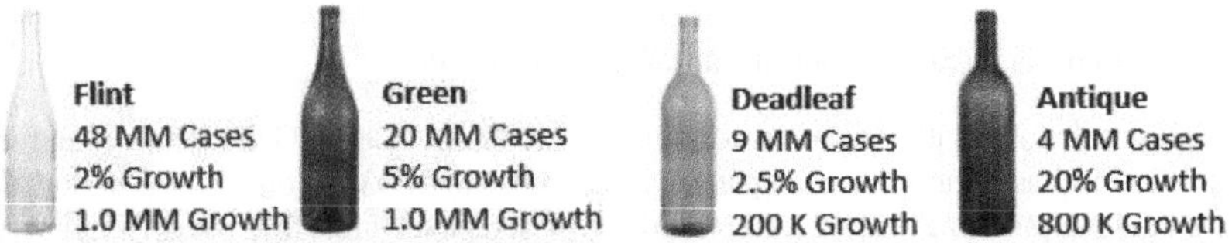

Figure 5. Glass Bottle Production Rates

We are adhering to Glass Sustainability principles. The benefits of glass, are its composition being made from natural raw materials, such as sand, limestone and soda ash, plus the fact that glass is non porous and impermeable (i.e., no chemical interactions), which makes it an ideal packaging material that maintains the quality and purity of its contents.

Just glass alone, is great for the products it contains, but Gallo Glass goes beyond by recycling the glass with each bottle being fifty percent (50%) recycled glass. Gallo Glass supports these measures with a dedicated recycling plant. Glass is endlessly recyclable unlike other packaging materials.

Furthermore, we reduce energy consumption by two to three percent (2% - 3%) for every ten percent (10%) of cullet utilized. Additionally, for each six (6) tons of cullet used, there is a reduction of CO_2 gases by one (1) ton of CO_2 gas. Figure 6, below, illustrates the Melting, Forming, and Inspection processes.

Figure 6. Our Glass Melting, Forming and Inspection

The Melting processes include four (4) oxygen fired furnaces, producing 500,000 tons of glass per year of flint, green, deadleaf green, and antique green bottles. The Forming processes include fourteen (14) forming machines, ninety five (95) mold sets, and one tousand two hundred (1,200) SKUs. The Cold End processes include forty three (43) inspection machines, robotic palletizing, that are capable of handling 2.5 million (2,500,000) bottles per day. A majority of the glass manufactured by us is supplied to its parent company, E&J Gallo Winery, that produces the brands Apothic Red, Louis M. Martini, Barefoot and New Amsterdam spirits. Figure 7, below, shows the Future Capacity Footprint.

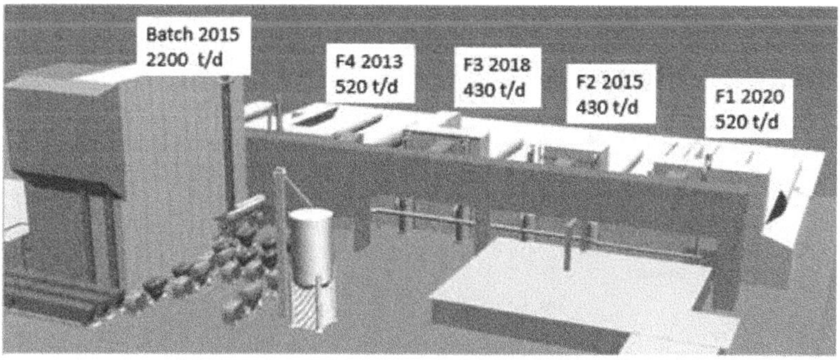

Figure 7. Our Future Capacity Footprint

Furnace 1 will be melting 520 tons of glass per day by 2020, utilizing one (1) single, and three (3) double gob machines producing an extra 700,000 cases of flint bottles. Furnace 2 has been melting 430 tons of glass as of 2015, utilizing two (2) double and one (1) triple gob machines, producing an extra 5,600,000 cases of flint and deadleaf green bottles. Furnace 3 will be melting 430 tons of glass of glass as of 2018, utilizing two (2) double and one (1) triple gob machines, producing an extra 5,600,000 cases of flint bottles. Furnace 4 has been melting 520 tons of glass as of 2013, utilizing four (4) double gob machines, producing an extra 500,000 cases of green bottles. Note that the new batch house has a 2,200 ton per day capacity.

FURNACE 1

The ***MBPC*** Chronology and Status for Furnace 1 is as follows:

- ✓ ***MBPC*** Installation (5-6/2014)
- ✓ ***MBPC*** Performance optimization (6-7/2014)
- ✓ ***MBPC*** Greater than three percent (3%) energy reduction
- ✓ ***MBPC*** Currently in use

The ***MBPC*** Chronology and Status for Forehearths 11, 12, 13, and 14 is as follows:

- ✓ ***MBPC*** Installation (2014)
- ✓ ***MBPC*** Performance optimization (2014)
- *x* ***MBPC*** Temperature stability improved, no improvement in gob weight variability. The MBPC initially had some gob weight variation due to aggressive temperature control. The MBPC was re-tuned to a slower response time that resulted in better gob weight control, but due to other limitations, the MBPC was not engaged.

Figure 8, below, shows the energy efficiency improvements on Furnace 1 in mmBTU/ton, normalized to fifty percent (50%) cullet.

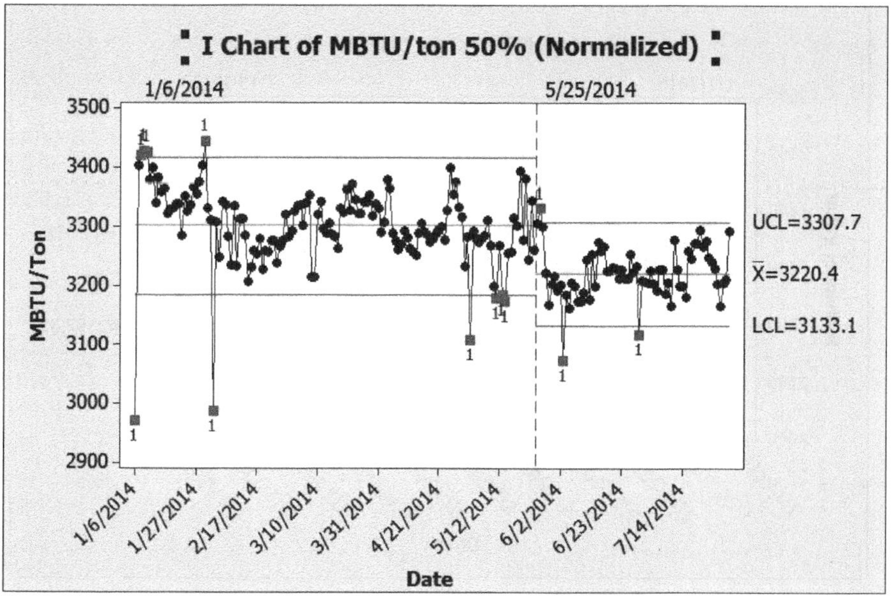

Figure 8. Energy Efficiency for Furnace 1

FURNACE 3

The ***MBPC*** Chronology and Status for Furnace 3 is as follows:

- ✓ ***MBPC*** Installation (7/2015)
- ✓ ***MBPC*** Performance optimization (10/2015)
- ✓ ***MBPC*** Greater than three percent (3%) energy reduction
- ✓ ***MBPC*** Currently in use

The ***MBPC*** Chronology and Status for Forehearths 31, 32 and 33 is as follows:

- ✓ ***MBPC*** Installation (7/2015)
- ✓ ***MBPC*** Performance optimization (10/2015)

Figure 9, below, shows the energy efficiency improvements on Furnace 3 in mmBTU/ton, normalized to fifty percent (50%) cullet.

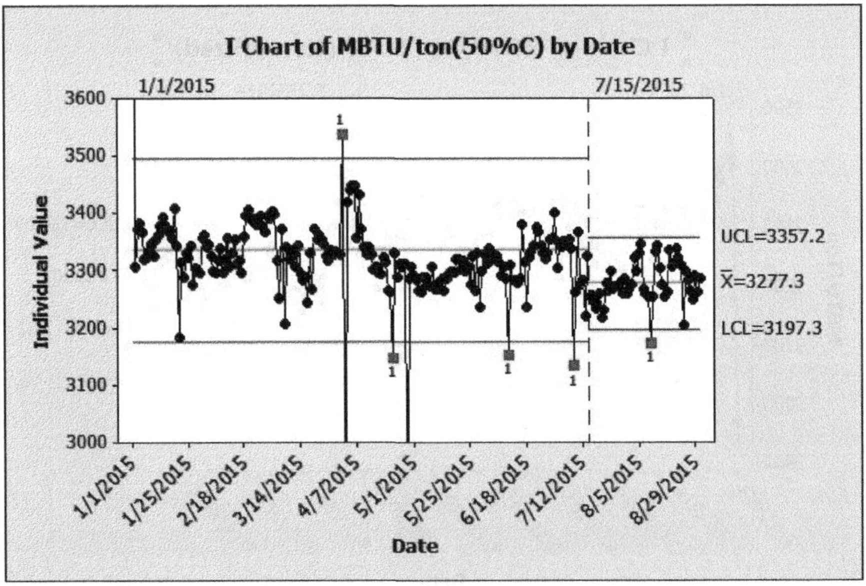

Figure 9. Energy Efficiency for Furnace 3

FURNACE 1 AND 3 OPERATING STRATEGIES

Standard Furnace Control:

- Well-up temperature is controlled by total gas set-point with higher priority than superstructure top temperature
- Electric boosting is only adjusted if the well-up glass temperature is more than 5°F from the target set-point
- Electric boosting is adjusted based on the actual pull rate

Emergency logic:

- If the superstructure top temperature exceeds a predefined maximum temperature, gas is reduced automatically

Job Change Logic

- Defined by the operator input (new pull rate, date and time of change)
- Superstructure top temperature is controlled by total gas set-point
- Well-up temperature is not controlled during the job change
- When tonnage is changed, electric boosting is adjusted based on user defined matrix of tonnage and electric boost (optional)

- When tonnage is changed, a new set-point for the well-up temperature is calculated based on user defined matrix of tonnage and well-up temperature (optional)

FURNACE 4

The ***MBPC*** Chronology and Status for Furnace 4 is as follows:

✓ ***MBPC*** Installation (7/2015)

✓ ***MBPC*** Performance optimization (10/2015, 12/2015)

x ***MBPC*** Energy savings has not been statistically proven

Electric boost reduction was offset by increase in gas usage

✓ ***MBPC*** Currently in use

FURNACE 4 OPERATING STRATEGIES

Standard Furnace Control:

- Well-up temperature is controlled by electric boosting only in full automatic control
- Superstructure top temperature is controlled by total gas set-point only Glass level is monitored and is an input into the control logic

Emergency logic:

- If the superstructure top temperature exceeds a predefined maximum temperature, gas is reduced automatically

Job Change Logic

- Identified automatically by batch charger speed and well-up temperature variation
- Superstructure top temperature is controlled by total gas set-point for a few hours after a job change
- Well-up temperature is not controlled during the job change and is controlled only after the job change has been completed
- MBPC system maintains temperature control even with a 5% change in furnace pull

Figure 10, below, shows that the energy efficiency has not been statistically proven on Furnace 4 in mmBTU/ton, normalized to fifty percent (50%) cullet.

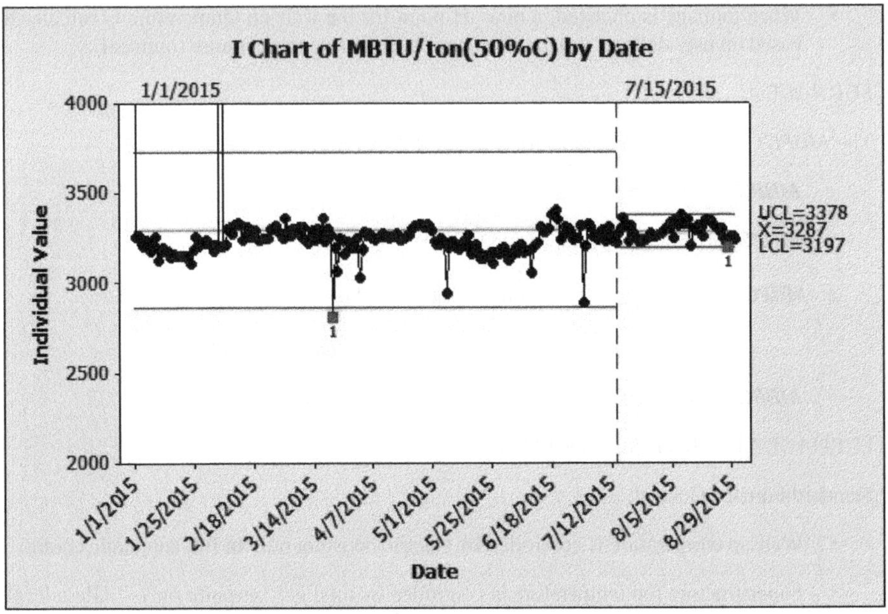

Figure 10. Energy Efficiency for Furnace 4

Figures 11 and 12, below, show that the stability on Furnace 4 improves with the MBPC system in operation as compared to when it was not in operation.

FURNACE 2

The ***MBPC*** Status for Furnace 2 is as follows:

- ✓ ***MBPC*** Installation planned (1Q/2017)
- ✓ Capital approval completed
- ✓ Energy rebate application in process

It is important to understand how temperature control and stability can really be used to save energy. Most glass furnaces are operated at an average glass temperature that is hot enough to prevent fining problems and seeds and blisters in the final product. The more the glass temperature varies, the hotter the average furnace glass melt temperature is chosen to prevent glass quality issues. With a controlled and stable glass temperature, the furnace manager is able to explore the limits for his glass production. By lowering the set-point of the operating temperature the furnace manager can save a considerable amount of energy. But this depends also on how the customer uses the tool most effectively.

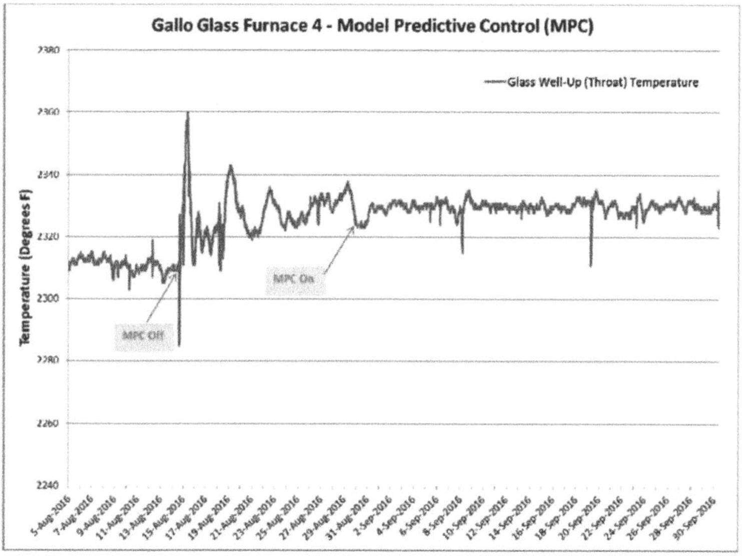

Figure 11. MBPC Operating Stability for Furnace 4

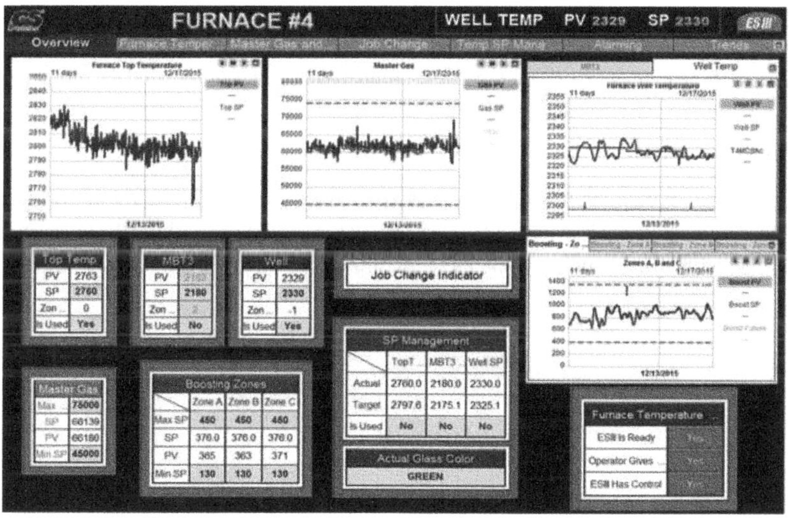

Figure 12. MBPC Operating Screen for Furnace 4

Figure 13, below, shows energy savings by temperature control and stability. The step down in the set-point of the glass temperature saves directly on average gas consumption by using the reduction in temperature variation to explore the actual limits of the melting process and to keep the desired glass quality.

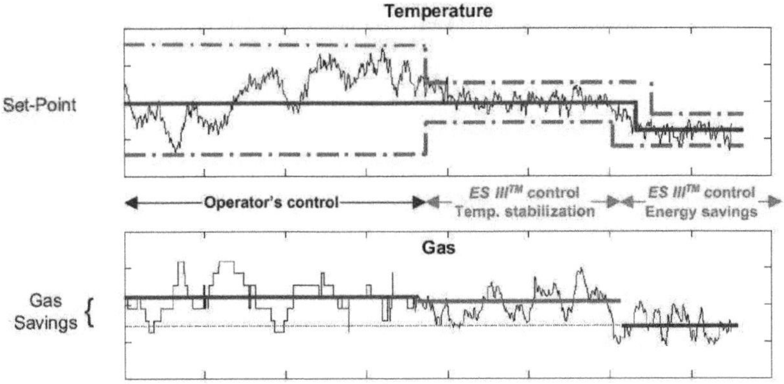

Figure 13. Set Point Changes for Energy Savings

CONCLUSIONS

There is still some work to do on the forehearth MBPC Systems, since there is no improvement in gob weight variability, even though there is improved temperature stability. Furnaces 1 and 3 have both yielded energy savings of greater than three percent (3%), with the melter operators preferring to utilize the advance control concepts offered by a MBPC system. Although the MBPC System has not yet shown a statistical energy savings for Furnace 4, the operators prefer to utilize the system because improved stability in the well-up temperature has led to improved forehearth control.

REFERENCES

[1] Josef Müller, Robert Bódi, Josef Chmelař: How to Make Glass Furnace Control Easier: Advanced Optimal Control by Expert System *ES III*^{TM}, Proceedings of the VIII. International seminar on mathematical simulation in glass melting, Velké Karlovice 2005, p. 200

SMART MANUFACTURING FOR CONTINUOUS, HIGH-TECHNOLOGY GLASS PRODUCTION

D. Kuhn and J. Ahrens
Corning Incorporated
Corning, NY 14831
USA

ABSTRACT

Smart manufacturing (SM) systems include data management and data analysis, which provide real-time actionable information that can be used to optimize and build system intelligence into manufacturing operations. The ability to utilize SM systems is essential for high-technology glass production. In liquid crystal display applications screen resolution, material strength, weight, and shape are all elements that require advanced manufacturing to produce industry leading glass products. Timely measurements, data analysis, process models, and process controls are key components of a SM framework as is the integration of these components through the automation of manufacturing tasks and the digitization of workflows that include both machines and human users. As with any continuous production process, these technologies are important for achieving high quality and high yields in glass manufacturing. The SM approach, the needs that it satisfies, and benefits that may be realized are described. Further, a SM framework that applies to continuous production of high-technology glass is described where advanced process controls are enabled through data integration, measurements, and modeling to improve process performance.

SMART MANUFACTURING AS AN ENABLER OF MARKET LEADING GLASS PRODUCTION/QUALITY

Smart manufacturing (SM) systems have been described by the National Institute of Standards and Technology (NIST) as "fully-integrated, collaborative manufacturing systems that respond in real time to meet changing demands and conditions in the factory, in the supply network, and in customer needs [1]." As glass manufacturing processes have grown and evolved, the need to meet ever increasing production and quality requirements has led to a proliferation of advanced process control and measurement equipment deployed online and near line. These computing devices have tended to follow (although lagging by a number of years) the computer industry performance trajectory (Moore's Law), with capability to collect and process information doubling at a rapid rate. This capability growth has been critical to allowing glass manufacturers to maintain sustained growth year over year. One good example of this growth trend is the LCD glass industry.

With increasing processing capability has come a concurrent increase in the ability of these systems to collect large amounts of data. This is fortuitous, because calculating the amount of data that must be collected by measurement systems looking for 10 micron sized features in LCD glass, the problem has ballooned from a 200 Petabyte challenge in 2000 to an 8 Exabyte one by 2014 based on shipped square footage alone. To this you must also add that with display density increasing (2x since 2011), these systems have the added requirement of finding smaller and smaller features. A reduction in feature size by 2× means a 4× increase in the amount of data that needs to be sifted to detect that feature.

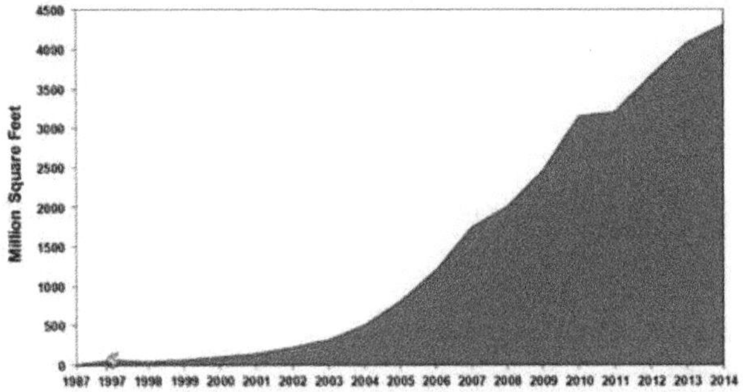

Figure 1. Historical LCD Shipment Growth[1]

Because of the high data volumes required for measurement and inspection activities in the LCD Display Glass and other high volume glass production markets, online and near line systems tend to employ a couple techniques to cope with throughput efficiently. The first technique leverages the fact that while an imaging system that is detecting small features on/in the glass may need very high resolution/bandwidth to detect these features… there are relatively few of these features occurring per square foot of glass. The resulting images are defined as “sparse images” because most of the image has no important information. This feature of many glass measurement problems allows for a reduction of the data analysis “Throughput” problem by one or more orders of magnitude without reducing the systems’ ability to measure the desired features. The next technique typically employed for these problems is to “analyze then dispose” of the raw data collected for the measurement. The historical cost of storing the raw data associated with these measurements has been high so other than engineering sampling, raw data was not stored. Measurements platforms are networked for aggregate statistic reporting but raw data is typically not stored, and even feature data is often only stored locally. These systems provide the needed inputs into a manufacturing site’s quality architecture… but have limited analytics potential without expensive manual data collection and contextualization steps.

It’s this second data reduction approach that we address and that SM seeks to change. One of the keystones of the evolution that SM seeks to make in manufacturing is adoption of the concepts of “Big Data” in the manufacturing ecosystem [2]. “Big Data” is a term used to capture a broad array of IT related activities and technologies, but relevant to this discussion, the component of “Big Data” most relevant at the measurements level is the notion of “Keep Everything” [3]. This is the notion that the classical method of extracting known business value from a system is no longer sufficient… we must now retain raw data in order to allow for the extraction of information that is unknown at the time/location of data capture. We summarize this change as a move from a “Storage is too expensive to keep things that we aren’t sure are valuable” mentality to a “Storage is cheap, we’ve spent substantially to collect this data and the value of the unknowns in it is too great to throw away” approach. The next logical step once you’ve deployed systems that are collecting large amounts of data for future analytics or for aggregation with other data sources in order to enable advanced techniques like Predictive Analytics across enterprises and connected supply chains.

[1] P. Bocko, Corning Inc., ARPA-E Workshop on Micro-PV, 8 May 2014.

This effort is clearly highly complex and multi-faceted, but to start to build skills and understanding of how to deliver on this vision, our effort is focusing on two specific areas, infrastructure and network architectures to support a "Big Data Capable" manufacturing line, and the data management systems needed to enable "Big Data Analytics" on stored data.

TOWARD A "BIG DATA CAPABLE" MANUFACTURING ENVIRONMENT

The ready availability of computers utilizing drives with large capacity (Terabyte level) has allowed measurements systems developers to readily integrate substantial data stores into both off the shelf and custom measurement systems. Typically, processed results and aggregate statistics from measurement systems are stored to Level 2 systems running standard Enterprise tools (SQL Server, Oracle, OSI-PI, etc), but these systems are not architected to be able to cope with the volumes of data associated with storing large amounts of raw process and measurements data. Historically this has created a semi-hidden second set of data for measurement systems... this is a high volume and rich data set of raw measurements and engineering data that is either discarded after immediate processing or stored locally on deployed measurement systems. This data often conforms to the current definition of "Big Data...", it includes Volume, comes in at Velocity, and with value based on its Variety (especially the unknown knowledge it likely contains, waiting to be discovered and leveraged to improve the manufacturing process).

There are a number of shortcomings to leveraging local system storage when taken in the context of a big data ecosystem.

1) Data Accessibility – Data stored on local measurement or process control systems will of course be distributed (multiple similar systems each with their own data), it will also be subject to the vendor or custom code environment that controls that system. The result is that often even basic data access (beyond ops support) is challenging, and in order to utilize the data for advanced analytics it is generally required to launch an effort to manually collect and align the data from multiple sources.
2) Drive Reliability – There will be substantial variability in the reliability of the storage system, it's up to the vendor/team that develops it to define its architecture. We've observed solutions for measurement systems ranging from single commodity drives (no fault tolerance) to software RAID (basic fault tolerance) and in some cases low cost RAID (either onboard or Network Attached). While these solutions provide progressively greater data reliability, none of them approach the reliability of even low cost Enterprise level storage systems.
3) System and connectivity performance – The ability of a networked system to access data stored locally on a measurement or process control system may be severely limited by either system constraints (external data access takes a back seat to operations) or the network connection (typically gigabit Ethernet).
4) Security – Online manufacturing systems should have limited or no connectivity outside of the local network. As the need to defend manufacturing operations from external threats increases, we're seeing an increase in access controls for manufacturing systems. Without the implementation of new approaches for accessing distributed manufacturing measurements data (in a secure manner), that data becomes essentially invisible and inaccessible to any users not standing at or near the system.

One of the key goals of SM as well as similar efforts worldwide ("Industry 4.0" in Europe, "Made in China 2025" in China, etc) seeks to address the shortcomings of current 'big data' class manufacturing solutions. While the effort is still young, the current focus is on leveraging low cost storage and compute technologies that are already well established in the Internet space. Cloud technologies such as Hadoop enable the use of commodity hardware yet provide Enterprise levels of data reliability. These systems offer a compromise between difficult to integrate and support local solutions and cost prohibitive (for local measurement systems use) Enterprise IT solutions.

SYSTEMS TO SUPPORT "BIG DATA ANALYTICS" – AN EXAMPLE

For a manufacturing infrastructure that has leveraged a "Big Data" compliant manufacturing data system, a number of new systems architectures are opened up to support advanced analytics. Over the near term, the approach of most interest utilizes the distributed nature of Cloud infrastructure to solve the problem of how to efficiently support analytics against large and geographically separated data sets. There are number of possible implementations that could provide a viable solution, one would involve a Hadoop driven multi-layer architecture. In this environment, Hadoop "Fog" compute clusters provide in-situ storage and compute capacity. These systems reside near line and provide robust IT compliant storage for big data generators. Additionally, these clusters provide compute capability to enable the first level of data aggregation. This first line of data reduction is critical to enabling advanced analytics because it reduces the data volume required to be transmitted to the enterprise cloud infrastructure (public or private) to a manageable amount. In this implementation, an analytics job would entail a distributed "Data Aggregation" job deployed to the geographically diverse fog nodes, followed by a data transfer from the fog nodes to the centralized cloud environment and finally an Analytics job run against a data set including all your data sources. The potential of such an environment to reduce a complex multi-week data collection and analysis project into a task that can be built and launched by a single Data Scientist sitting at his/her desk could reduce the cost of exploratory analytics jobs to the point that they become routine, dramatically increasing the value of the data being collected from measurement systems.

PROCESS MODELING, CONTROLS, AND SYSTEM MONITORING AS A PART OF SM

One of the chief aims of SM is to develop models and use them to make actionable decisions impacting process operations [4]. The actionable decision making agents can be human, including operators, engineers, management, and supply chain, or they can be automated systems in the form of process controls. The SM approach addresses the challenges and costs associated with automated systems and controls spanning multiple layers of operations. At the micro-level, real-time control systems are managing the process on time scales from sub-second to minutes. The meso-layer contains systems operating from hours to days. These systems will include supervisory controls, process monitoring, fault detection, and maintenance. A macro layer system operates on the order of weeks to months or longer and handles supply chain decisions and operational scheduling. We briefly describe process control and predictive analytics in SM for the first two layers next.

For real-time control situated at the micro-layer, process models can be incorporated to improve operations. Figure 2 shows a simple block diagram of a closed-loop process control system that makes ideal use of process models. High fidelity models, which are often based on finite element approaches, contain detailed information that may predict process performance or determine proper operating conditions. These models contain detailed, spatially distributed information but are not suited to real-time control due to their complexity. One approach is to use Reduced Order Models (ROM) that are capable of simulating in real-time but are accurate enough for process control design and execution. These reduced order models can be updated by the high fidelity models to account for changes in operating conditions or improvements to the model's predictive capability through the use of data collection and predictive analytics. This is an important approach for precision glass production where applications to thermal control in the glass melt and annealing processes can have a large impact on the final glass quality. With a "fog-based" data management and control system infrastructure, the real-time systems can be executed in the micro-layer whereas the high fidelity models may reside in the enterprise cloud layer. This has benefits for the support of systems running in plants across the globe as will be discussed below.

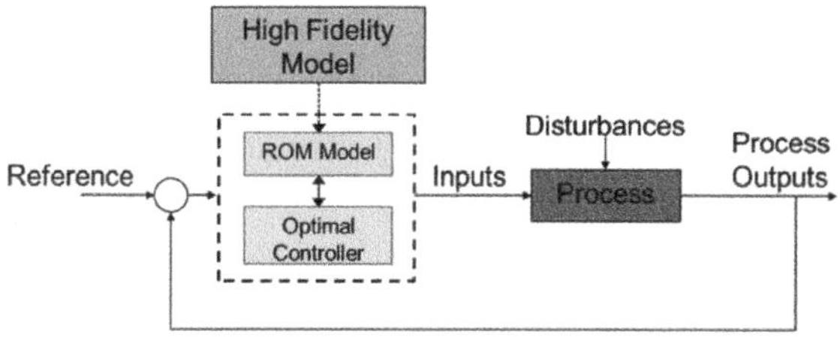

Figure 2. Model-based control block diagram showing micro layer process controls. Bringing data and models to the process to make actionable decisions is an integral part of SM. An optimal controller which makes use of a Reduced order model (ROM) is used for real-time process operation. High-fidelity models are used to update the reduced models based on new operating conditions.

Meso-layer systems contain significant potential for building manufacturing intelligence into process operations for glass production. Here systems that facilitate extensive data collection, visualization, and analysis can be built for a process to maintain optimal, or near optimal execution. As mentioned previously, data driven methods can be implemented at this layer to handle process monitoring and fault detection systems (for a good survey of industrial process monitoring applications see [5]). Figure 3 shows a diagram of process monitoring and fault prediction. These systems include equipment monitoring applications to automatically identify faults. Real-time control system performance monitoring is also included. Changes in process disturbances, equipment degradation, or changes in operating conditions can cause the controls to perform sub-optimally. Performance metrics, such as the predictive capability of real-time models, can be tracked and the controls and models can be updated to account for new conditions. For precision glass production, tight tolerances on product quality places a strong emphasis on being able to observe those key process variables that impact product quality. Building data-based models that take these process variables as an input and predict quality performance as an output is an enormous advantage for operation. Successful application of such an approach can save costly quality inspection. With sufficient data for accurate modeling and validation, process monitoring applications have the potential for prediction of process and quality problems. Short of the ability to anticipate process behavior, monitoring to quickly flag and address process, quality, or equipment issues when they occur still has the potential for significant cost savings in glass manufacturing.

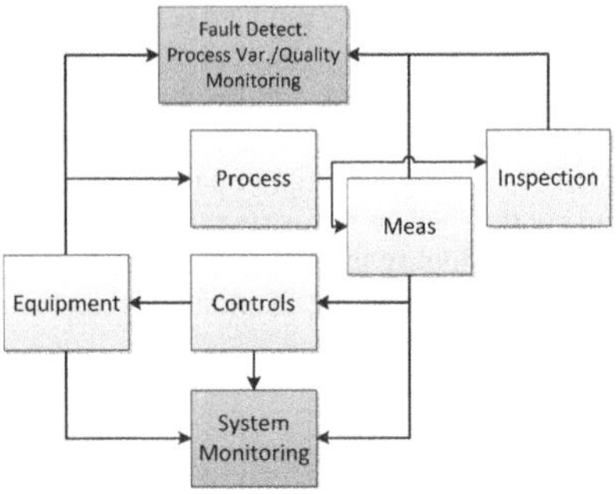

Figure 3. Diagram Illustrating Process Monitoring and Fault Prediction Technology. Process equipment, process variables, controls systems, and product quality can be monitored and managed through predictive analytics (see [5] for more detail on industrial process monitoring).

Developing the models and controls and implementing them on cost competitive infrastructure is one of the goals of SM. Long-term support of model-based systems presents challenges for organizations with limited technical personnel and this presents additional opportunities for a SM framework. Real-time and model-based control systems are integral to SM and enhance glass manufacturing production. Once these systems are deployed their maintenance is critical to sustaining their performance advantages over the lifetime of the process. As mentioned, changes in operating conditions, machine and equipment wear, and changes in disturbance characteristics can cause these systems to run non-optimally. This is a very important point that must be considered in a system deployment plan. Non-optimal or poor performance may very likely force operations to discontinue the use of model-based control systems in favor of ad-hoc manual approaches. This is especially the case if the plant lacks personnel with the proper technical background for model-based system maintenance. However, networked systems and integrated data can be used to support multiple processes in a plant and across multiple plants from a remote location.

One concept is to use a centralized system to monitor advanced measurements and control systems running in the field. This system would collect data from disparate sources to assess process capability and issue system updates to the plants. For instance, an engineering support application would monitor data from quality inspection systems as well as historian data that collect time series information on process operation. In addition, the system can collect process log data that may be in the form of text based information which contains key observations by process engineers and operators that details special events occurring on a process providing proper context for the other sources of data. The support application may interface with several systems running on the plant floor such as the real-time process control systems (Programmable Logic Controllers – PLC, or Industrial Personal Computers - IPC), the supervisory control and data acquisition system (SCADA), as well as a remote scientific computing environment (see

Figure 4). The functions of the application include performance monitoring capability to assess product quality and control system health. System dashboards can be used to visualize the performance of lines running in plants around the world, quickly assess system health, and flag for systems that need updating. The application can also include algorithms to detect system faults. Long-term data can be used to assess the predictive capability of the models running on the process. System faults can be issued if the model prediction begins to deviate beyond an acceptable threshold. The engineering application would also include the capability to derive reduced order models from high fidelity models and simulate them to evaluate their suitability for process control.

The remote engineering support platform can be used by subject matter experts who are trained to be able to interpret the system health indicators, determine the required model adjustments, and take steps to update the process model. As they are remote to the plants, they can provide centralized support for all systems running in the field. Therefore, only a relative few subject matter experts are required to keep deployed control systems running optimally. The process for detecting suboptimal system performance and execute a model update can be carried out through an automated procedure or workflow.

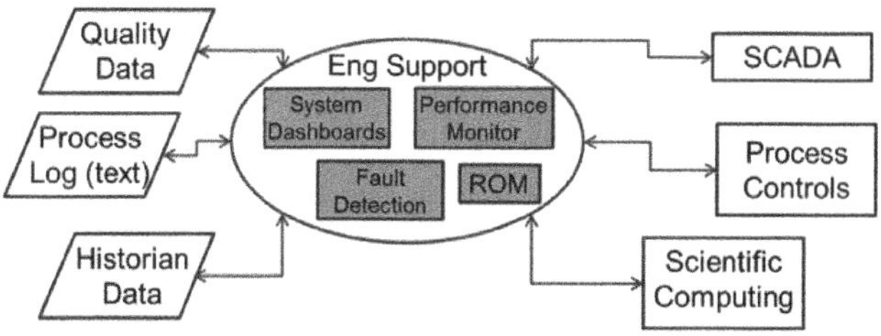

Figure 4. An Engineering Support Platform as part of a SM Framework - This can facilitate the management of model-based control tools remotely.

Figure 5 shows a sketch of how a process monitoring and support workflow might be executed. A remote application can monitor processes worldwide and flag for non-optimal performance. Data can be collected to adjust process models and a job can be kicked off to a scientific computing environment sitting in the cloud. A modeler can execute the job, which may take the form of lengthy Finite Element Model (FEM) simulations, and send the resulting model information back to the remote support application. The support engineer can run simulations and send the updated process controls to the plant where the plant engineer can approve the change and get back to running at optimal conditions quickly. This is an example where remote support can keep processes running efficiently across the globe.

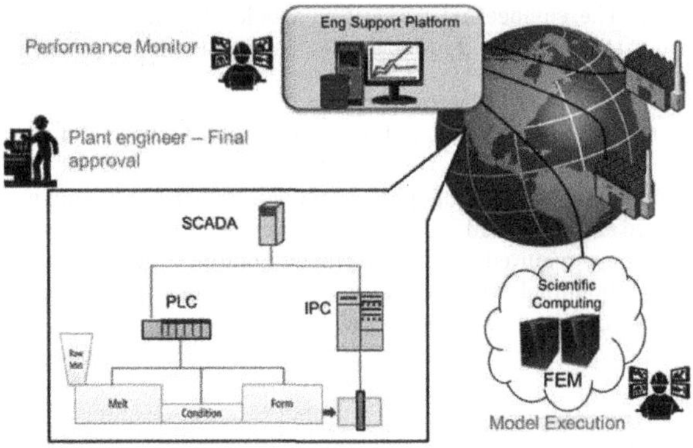

Figure 5. Diagram Illustrating the Workflow for an Engineering Support Application for Precision Glass Production Built on a SM Platform.

CONCLUSION

As Moore's Law has driven the orders of magnitude increases in computation and storage availability, this additional power has quickly been leveraged in some technology areas (Cloud Computing for instance) but has seen slower adoption in manufacturing. It is only over the last few years that we've started to see enterprises and industry groups seeking to enable manufacturing to leverage these enhanced technologies to shorten development cycles, improve efficiency and quality… and through these – competitiveness. These are large efforts and are receiving national support (US – CESMII, DMDII, Europe – Industry 4.0, China – Made in China 2025), and we are likely to see the impact of these initiatives in the Glass Industry emerging in the near to mid-term. Now is a good time for glass manufacturers to start to review these efforts to determine how they might want to participate as well as when it might make the most sense for their own manufacturing operation to launch a SM effort.

Advanced modeling and controls can have a big impact on process operation but they can come with significant costs. A SM platform has the potential for these techniques to be designed and implemented faster, thus reducing development costs. An open infrastructure as well as Software as a Service approach lowers the barriers to entry for glass manufacturers. Publicly available software solutions that address process operational challenges that have similarity across the industry can be adapted and implemented on a low cost platform without having to redevelop the solution from scratch.

Glass manufacturing processes producing disparate sources of data from batching, melting, refining, forming and annealing, finishing, inspection, and packaging generate significant sources of data, particularly for precision applications where inspection may require fine spatial resolution over a large area. In addition, for multiple production lines running in multiple facilities spanning the globe there are opportunities for aggregation of broad data sources to produce predictive models for the design of actionable decision making systems. The SM infrastructure provides a cost effective means for making this realizable.

REFERENCES

[1] "CTMC's Guide to Smart Manufacturing." http://offers.cmtc.com/smart-manufacturing-ebook, 2016, California Manufacturing Technology Consulting (CTMC) and the Smart Manufacturing Leadership Coalition (SMLC).

[2] NIST Real-Time Data Analytics for Smart Manufacturing Systems Project - https://www.nist.gov/programs-projects/real-time-data-analytics-smart-manufacturing-systems-project.

[3] HPE Business Insights - https://www.hpe.com/h30683/us/en/strategic-business-insights/c/big-data/business-intelligence/keep-everything--the-new-data-imperative.html.

[4] J. F. Davis, J. Wetzel, and B. Graybill, Smart Manufacturing, "Real-Time Networked Information Workflows and Enterprise Performance," in *Proceedings of the 3rd International Conference on Sustainable Chemical Product and Process Engineering,* (2013).

[5] J. S. Qin, "Survey on Data-Driven Industrial Process Monitoring and Diagnosis," Annual Reviews in Control, 36, 220-234, 2012.

OPTIMAL CONTROL STRATEGY FOR PREDICTIVE COMPENSATION OF GAS QUALITY FLUCTUATIONS IN GLASS MELTING FURNACES

P. Hemmann
STG Combustion Control GmbH & Co
KG Cottbus
Germany

ABSTRACT

German glass Industry is facing increasing problems – better call it "challenge" – born from fluctuating composition of natural gas. For cost optimization, gas suppliers are frequently switching between different sources of natural gas, taking bio gas from local sources too. In near future we may expect hydrogen from windmills as well. This paper presents a strategy for balancing different sources of gases used and their quality for glass melting.

INTRODUCTION

Mainly focused on the needs of millions of private consumers, gas suppliers are obliged to keep within limits of the WOBBE index only – which is not sufficient for stable operation of industrial furnaces. Even with nearly same WOBBE index, different gases come with different calorific values, which come with different demands for oxygen or air per Nm^3, and they come with different flame radiation characteristics, and therefore with different thermal efficiencies. It will take not only the reliable measurement of gas quality, but it will take efficient and predictive control strategies to deal with such challenges. Figure 1 shows variations in concentrations of different gases present in select H-gases.

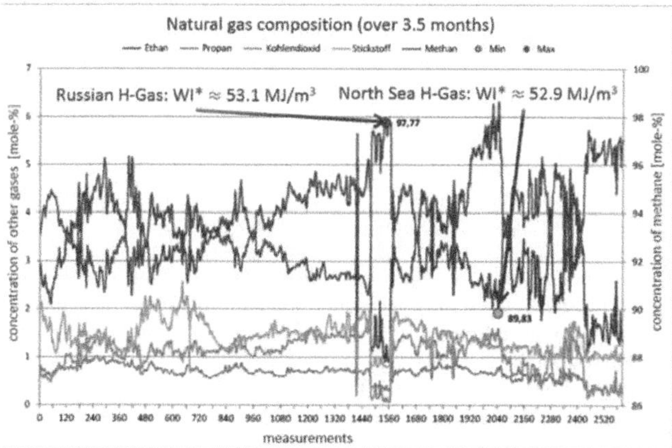

Figure 1. Concentration of Other Gases Present in Select H-gases

Major control targets are: 1) keep constant melting temperatures despite fluctuating gas quality, 2) keep constant combustion and redox conditions – for low NOx- and SOx emissions, 3) keep constant heat emission of flames and its distribution in furnace chamber, and 4) keep constant energy remaining in the furnace better than keeping energy input only. On the other hand, a control technology giving predictive response to gas quality fluctuations should be robust, *with as low as possible sensibility against accuracy of gas quality measurement.* The key for successful control is the optimal control strategy, rather than requirements for extremely high accuracy of gas quality measurements. Finally, it is all about seemingly simple questions: What is the right gas flow, and eventual secondary fuel flow too? What is the distribution over the melter ports? What is the right oxygen- or combustion air flow? And What is the right furnace pressure?

OLDEST FORM OF TRADITIONAL CONTROL

This is still from days of single loop controllers. Temperature Control requires a certain set point of the combustion air – (if the temperature control system would be working at all) - and the set point of the gas follows the actual value of the combustion air in a fixed ratio. This is safety oriented, giving the amount of the gas only for the amount of the actual combustion air. Surprisingly such an old fashioned control strategy may be close to a simple but effective response to gas quality fluctuations. Supposed You may fix the right ratio fitting to the modified fuel composition. But in reality, the gas quality fluctuation is not the only impact to compensate: Such a control response is extremely inflexible to compensate for the fluctuations of air ingress. By way of a theoretical expectation, any compensation of air ingress into the furnace by a modified combustion air to fuel ratio may work via increased or decreased gas flow, and the temperature controllers response to increased or decreased furnace temperatures will finally give the necessary adjustment of combustion air – this would take hours for a control action that should be done in minutes. And last not least, such an old fashioned control strategy will not have a good answer to compensate for the modified thermal efficiency.

CHANGING THE ORDER OF CONTROL LOOPS

This will give much greater flexibility. Temperature Control requires a certain set point of fuel gas or of fuel gas energy. This is flexible for a dual fuel operation as well, for such an example of a mix fire of gas and oil. The combustion air set point or oxygen set point will follow in a flexible ratio, which is subject to modification in order to compensate for the fluctuating stoichiometric ratio of fuel and uncontrolled air ingress - just in time. For PLC-based control, safety requirements may be fulfilled by Cross-Limit-Control – where set point of fuel is limited in the case of insufficient air flow. Such a control strategy is quite common, but it is not robust enough for energy control: Gas quality measurement will affect this sequence of control loops at least twice: first for the transition from temperature to energy and gas flow, and again for the adjustment of the ratio as a function of the gas composition.

For a full and comprehensive predictive compensation of gas quality fluctuations, one has to take care for the following problems (shown in Figure 2):

1) Density correction of the gas flow measurement, which will be different for differential pressure based measurement upon example mass flow meter measurement
2) Calorific value correction for the gas flow set point, in order to compensate for constant energy input
3) Correct different thermal efficiency born from air flow and flue gas flow enthalpies as a function of mass flows and temperatures
4) Correct different thermal efficiency born from different flame emissivity, as a result of carbon based flame radiation, which is higher for higher percentage of higher hydrocarbons
5) Correct different thermal efficiency born from different flame impulse, affecting the speed of combustion, mainly as a function of gas injector design
6) Correct set point for oxygen- or combustion air flow as a function of stoichiometric ratio of given gas composition, modified by actual air ingress as a function of furnace pressure and eventually ambient temperature

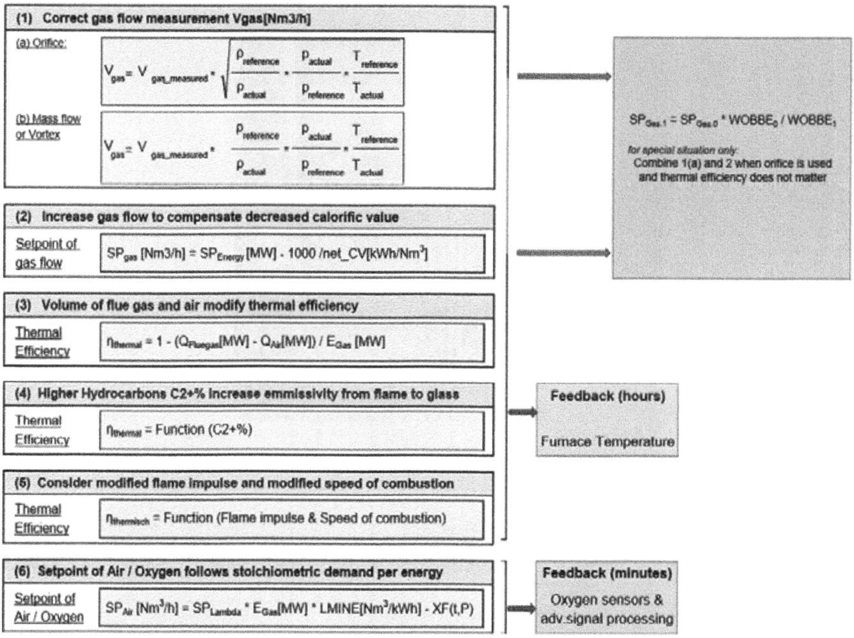

Figure 2. What Needs to be Done to Compensate Fluctuating Gas Quality

There are different solutions available for the measurement of gas quality. Starting from a "simple" calorimeter, providing calorific value (shown in Figure 3), WOBBE index and density only, coming with easy and robust self - calibration but considering gas components such as hydrogen and CO as well – up to different models of more sophisticated gas chromatographs, providing a full set of natural gas composition data, but requiring high efforts for maintenance and calibration, where only the most advanced models include information about hydrogen and CO. In general, the most expensive solution is not automatically the best solution to fit for the technological needs of a reliable furnace control.

In any case, the PLC embedded combustion calculation (shown in Figure 4) is the interface between gas quality measurement and subsequent control algorithms. This function block works with either a full set of gas composition data or with limited information, using a typical gas composition – the "working point"- modified by actual input of gas quality measurement.

Figure 3. A Simple Calorimeter

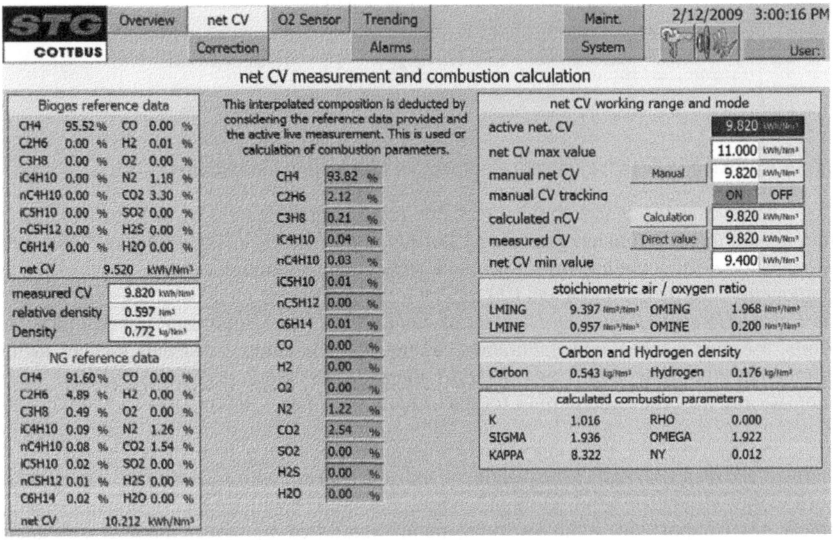

Figure 4. PLC Embedded Combustion Calculation

The combustion calculation provides all of the required information for subsequent controls:

- Gas density – for density correction of gas flow measurement
- Net calorific value – for energy correction of gas flow set points
- Stoichiometric ratio LMIN for the demand of combustion air per the volume of gas Stoichiometric ratio LMINE for the demand of combustion air per the unit of gas energy
- For oxyfuel combustion OMIN or OMINE for the demand of oxygen accordingly

- 6 dimensionless parameters K, SIGMA, OMEGA, KAPPA, NY, RHO for the advanced signal processing of oxygen sensors to get the actual value of Lambda from sensor signals
- Percentages of carbon and hydrogen in the fuel gas – to consider the impact on flame radiation
- Parameters for on-line energy balance as AMINE – stoichiometric ratio of flue gas – and $Cp._{gas}$, $Cp._{flue}$, specific heat capacities of gas and flue gas.

There may always be some minor inaccuracies due to incomplete measurement of gas quality, but what is really essential is: to work with a consistent data base, where all of the data for gas flow control, for air or oxygen flow control, for oxygen sensor signal processing and for calculations of the energy balance and of the thermal efficiency are logically fitting together. PLC embedded combustion calculation can just provide this consistent data base. Necessary corrections and adjustments are described in detail below.

DENSITY CORRECTION OF THE GAS FLOW MEASUREMENT

Each kind of gas flow measurement is typically based on a fix gas density reference, but in reality, the gas density will be different from this reference. Therefore the gas flow measurement needs a correction to get the real actual flow. Such a correction may come to a not neglectable range of up to 1.2%. The actual correction depends on the type of gas flow measurement, means: whether it is born from a differential pressure – such as an orifice or venturi type – or not from a differential pressure – so for mass flow meters. In our table it comes combined with the temperature- and pressure-compensation. Independently from density correction, it takes a modification of the gas flow set point to keep energy flow constant. Such a gas flow set point correction comes with the actual net calorific value CV.net of the fuel gas (see formula in Figure 2). Alternatively, some users advocate application of the WOBBE index for the gas flow correction. One might combine the density correction of the gas flow measurement and the CV.net correction of the gas flow set point in one step, but this works for differential pressure based gas flow measurement ONLY and one looses the base for energy control of dual fuel combustion, as for example gas and fuel oil (there is no WOBBE index for oil, and such thing like "WOBBE flow" to be compared with energy flow from fuel oil is simply not existing and physically not correct). In addition, the adjustment of combustion air or oxygen we need the net calorific value any way

Keeping WOBBE index constant means, to send at same pressure through the same nozzle size the same energy, which is essential for millions of private consumers. This is common for fore hearth controls, whereby diluting the gas with air to keep a constant WOBBE index, since typically fore hearth burners are running with only pressure-controlled mixing of combustion air and gas. The same method is not applicable for a melting furnace: it would increase NOx formation due to early mixing of the gas and air at high temperatures. Building an energy furnace control on the net calorific value application makes things more logic and keeps doors open for an energy based duel fuel mixture fire control system and for energy based combustion air or oxygen flow control.

When we are looking for the influence of gas quality on combustion air demand, we find for a majority of natural gases, that the stoichiometric ratio LMIN for air-to-gas mix is following the net calorific value CV.net in nearly the exact proportion. So for a majority of applications the energy based stoichiometric ratio LMINE, defined as:

LMINE [Nm3.air/kWh.gas] = LMIN [Nm^3.air/Nm^3.gas] / CV.net [kWh.gas/Nm^3.gas]

A constant value of LMINE = 0,956...0,958 Nm^3/kWh, may be used. Trying the same with a number like LMIN/WOBBE we get a number showing typically a 10 times higher variation than LMINE=LMIN/CV.net. Figure 5 shows an example.

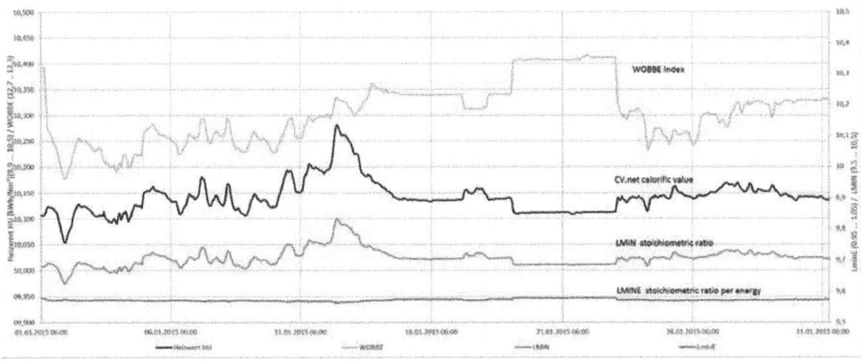

Figure 5. An Example Over a Full Month, where CV.net, WOBBE and LMIN are varying more or less strongly, but the energy based stoichiometric ratio stays in a band of LMINE = 0.9571 - 0.9578 Demand for combustion air or oxygen follows energy flow in so close linearity, that this may be used for a robust control loop for combustion air:

$$SP_{.air} = SP_{.Lambda} * LMINE * MAXIMUM(SP_{.Energy}, PV_{.Energy}) - XFM(pressure, time)$$

where $SP_{.air}$ = the set point of air, $SP_{.Lambda}$ = the technologically required surplus of air in dimensionless form, $SP_{.Energy}$ and $PV_{.Energy}$ = the set point and actual value of energy input XFM (pressure, time) represents a model of air ingress as a periodic function of furnace pressure and of actual time in reversal period.

Instead of a variable air-gas ratio this works with a variable supplement XF(pressure, time). We call XF "uncontrolled air", considering that XF may be positive or negative, for example by loss of combustion air due to leaky reversal damper. Figure 6 shows typical sources of uncontrolled air XF.

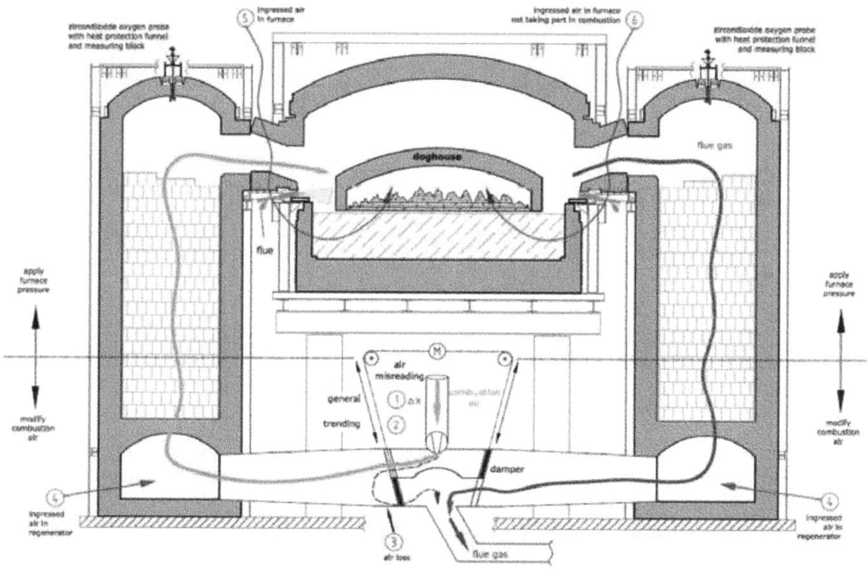

Figure 6. Typical Sources of Uncontrolled Air XF

Actual value for XF is a result of flue gas analysis in the regenerator crown. Advanced signal processing of oxygen sensors gives us O2%, CO[ppm] and the actual lambda value as:

$$\lambda = 1 + \frac{0{,}2094 * (\kappa + \omega + Ge) - 0{,}7906 * \psi}{\sigma} * \frac{[O_2\%]}{20{,}94\% - [O_2\%]} - \frac{\psi}{\sigma}$$

where Ge = relation of CO_2 from batch against CO_2 from combustion, ψ = relation of CO against CO_2 – to get from sensor signals like O_2%, and σ, ω, K represent dimensionless fuel parameters according to BOIE. So we get actual value XF for uncontrolled air as a difference of the totally indicated air minus actual value of controlled combustion air $PV_{\cdot Comb.Air}$, as shown below:

$$XF = \lambda * LMINE * PV_{\cdot Energy} - PV_{\cdot Comb.Air}$$

It is easy to understand, that XF also acts like kind of "ragpicker", collecting inaccuracies of all related measurements. Finally, based on the turbulent fluctuating values of XF, regression analysis provides the *XFM(pressure, time) as a periodic function of the furnace pressure and time, giving a model of the disturbance variable of the combustion air control* - the key for predictive control of optimal combustion air. Figure 7 shows an example.

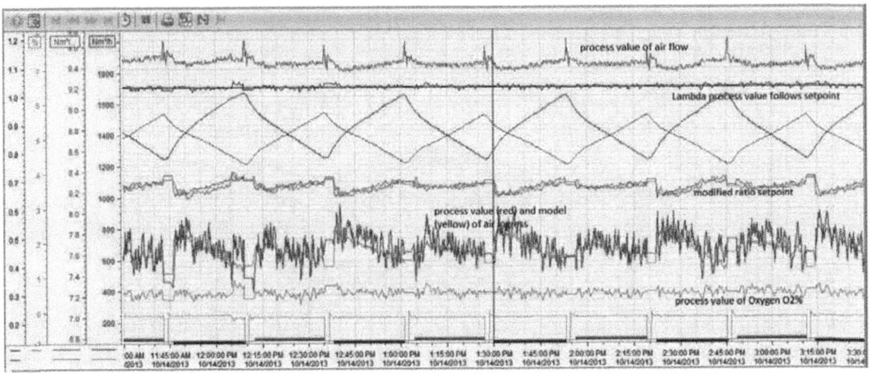

Figure 7. An Example of XF and Automatic Control of Lambda

For the above mentioned steps of density correction of the gas flow measurement, CV.net correction of the gas setpoint, and combustion air following the energy with an automatic Lambda-compensation of uncontrolled air. The basic principles are quite clear and the described functions are successfully in use. The next logical step forward will be: Control the amount of energy which REMAINS in the furnace instead of controlling the energy INPUT only. Such approach will make thermal efficiency η_{therm} (as a function of gas quality) an additional subject to automatic control. So we have to take care for the following:

- Different amount of flue gas and combustion air and their temperatures will influence the thermal efficiency, which needs to be compensated
- Higher percentage of higher hydrocarbons – means higher percentage of carbon against hydrogen – will increase the emissivity of flames and therefore will increase the thermal efficiency, which needs to be compensated
- Different amount of gas flow will result in a different flame impulse and a different speed of combustion, which needs to be compensated

Without a robust theoretical base for a practical control application, actually these aspects have to be observed and considered on an empirical base, taking thermal efficiency as a function of CV.net and of percentage of carbon in the flame.

SUMMARY

Figure 8 shows a summary of all aspects explained above along with the optimal control strategy:

- Temperature Controller provides set point control for energy-demand in terms of [MW]
- Calculate thermal efficiency η_{therm} as a function of carbon density and CV.net
- Setpoint for Energy input = Energy demand / η_{therm}
- Calculate setpoint of Combustion air

$$SP._{Comb.air} = SP._{Lambda} * LMINE * MAX(SP._{Energy.input}, PV._{Energy.input}) - XFM(pressure, time)$$

- Get Cross-limitation of gas for insufficient air

$$CLIM = MIN\{1, (1+tolerance)*PV._{Comb.air}/SP._{Comb.air}\}$$

- Calculate the set point for fuel gas flow with cross-limitation for insufficient air

- $SP.Gas = CLIM * SP._{Energy.input} / CV.net$
- Consider density correction for gas flow control

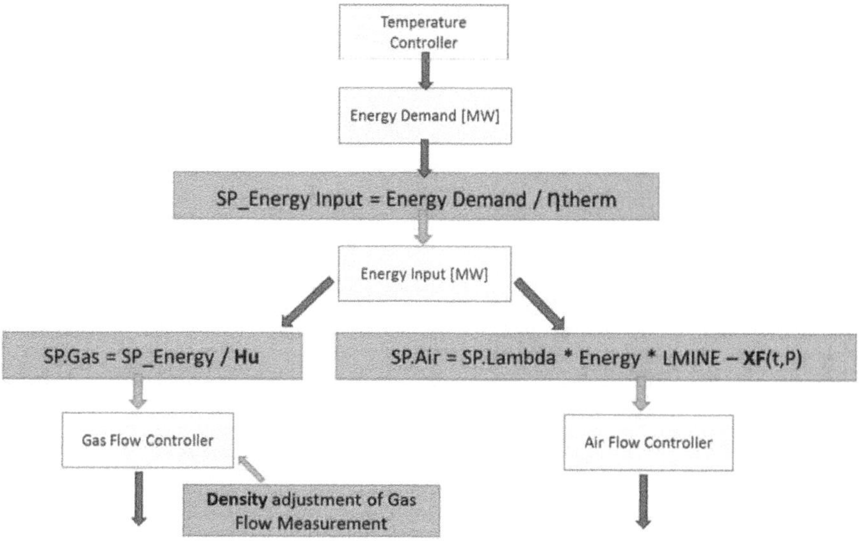

Figure 8. Energy Based Control Strategy to Compensate Fluctuating Gas Quality (patent application in progress)

The same strategy can be easily applied to a dual fuel mix fire control of gas & oil or gas & pet coke. Overall benefit of deploying this strategy is the stability that gives higher melting rate and less emissions, consuming less energy.

Energy

NEW FURNACE DESIGN SOLUTION BREAKS THE 3 GIGAJOULES PER TON BENCHMARK LIMIT

M. Lindig-Nikolaus
Sorg GmbH
Germany

ABSTRACT

Glass melting furnaces are considered to be high energy consuming industrial fabrication facilities. Improvements of furnace designs to improve energy savings are a continuous demand of the glass manufacturing industry. Reductions of greenhouse gas emissions to minimize effects on global warming are creating additional demands for further investigations to enable manufacturers to comply with future greenhouse gas limitations. There are many approaches and inventions targeting improved energy efficiency. Most ideas are never implemented full scale since visible improvement is not achievable with the proposed solutions. An example of this could be a proposal that improves melting rate, but does not necessarily result in more energy efficient operation. This paper presents a new furnace design solution.

CHALLENGES AND INNOVATIONS IN THE PAST

Continuous improvements in energy efficiency are a requirement to compete in the global market. The first oil crisis in 1973 and the second crisis in 1979 prompted the first serious activity in the area of improving energy efficiency. A significant change in boundary conditions for the industry came with the arrival of the first environmental regulations. In Germany the first decree for emission limits came into force in 1986. NOx emission limits were set for the glass industry depending on kind of glass composition and application of fining agents. In 1987 the SORG company invented a low emitting, energy efficient furnace solution, the LoNOx melter. Emitting less than 400mg/Nm^3 NOx the furnace operated well under the official NOX limits. The energy efficiency of that original LoNOx melter can still compete with the today's European benchmark, summarized in the BREF report.

The original LoNOx melter was equipped with a flue gas heat recovery system to preheat the cullet charged into this furnace. It was one of the first installations of I its kind, and was implemented at about the same time the first public container glass collection system was established by the glass industry.

Continuous improvements of this type of energy recovery solution gave rise to the latest invention, a special furnace design with a complete solution for batch preheating and a specially designed batch charging machine (Figure 1).

In 1981, we have patented another approach in melting improvement, titled the Deep Refiner, a solution specifically for end port fired furnaces allowing an increase in specific pull and improvements in fining conditions. This furnace design feature has since been established worldwide in the majority of these kind of furnaces (Figure 2).

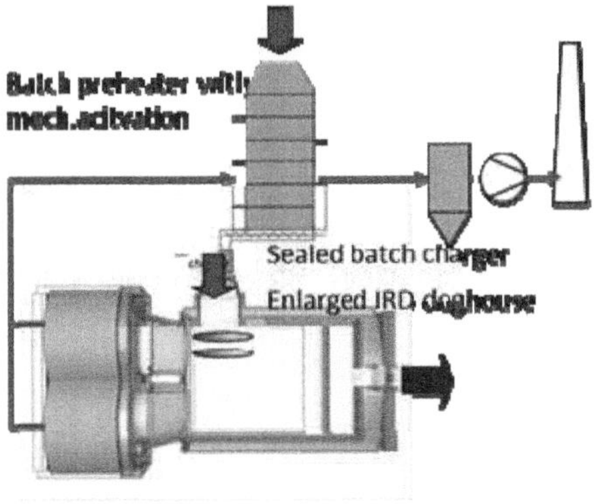

Figure 1. Concept with the Improved Features for Batch Heat up, Handling, and Pre-melting Zone

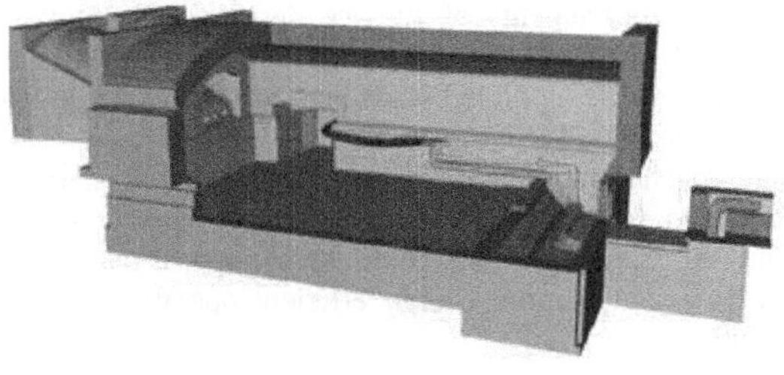

Figure 2. Deep Refiner in CFD Modelling with the Critical Path. The front end deep section allows melting capacity increase.

Another example of a heat loss reduction solution is changing the furnace doghouse design. Historically the charging pocket was almost open, causing significant radiation heat losses and dust pollution in the charging area. In 1996 SORG invented a completely closed doghouse with special charging machines using screws or vibrating tubes for batch delivery into the furnace.

Emission limits for greenhouse gases, were typically achieved by applying primary measures like near stoichiometric firing or asymmetrical positioning of the burner nozzles in the port. Even with improved firing NOx emissions cannot be reduced significantly below 800mg/Nm^3 (which corresponds to 1,8lb/USt for a container glass furnace).Trying to achieve today's emission limits using operating conditions of the past will result in lower specific pull of melters and clear loss of productivity (Figure 3).

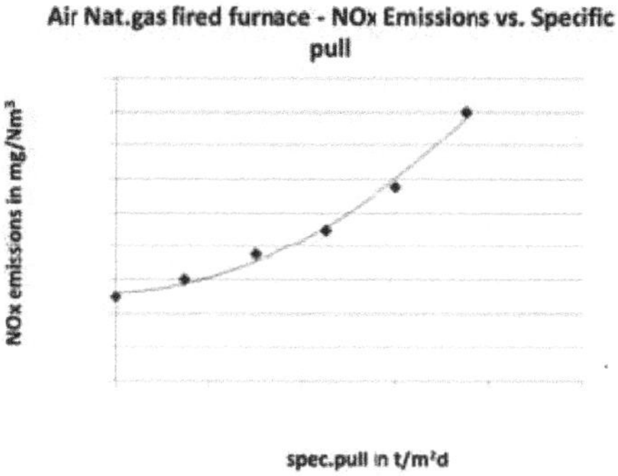

Figure 3. Evaluation of the Influence of Primary Measures against NOx on Productivity

FUTURE CHALLENGES

Environmental legislation in the past was almost entirely focused on pollutants and their concentrations, such as NO_x, SO_x, HCl, and dust. Global warming and global concerns about this phenomenon have more recently directed public attention to CO_2 emissions. Not only pollution but combustion gases in general have become an issue for the global industry.

In addition, allowable emission limits are being reconsidered within ever shorter intervals. What might be feasible today by adjusting primary measures to decrease pollution will probably not be sufficient in the future, and instead, secondary measures for pollution abatement will be required. The new and very efficient flue gas cleaning facilities now available allow high melter performance without the impacts from having to change primary measures (Figure 4).

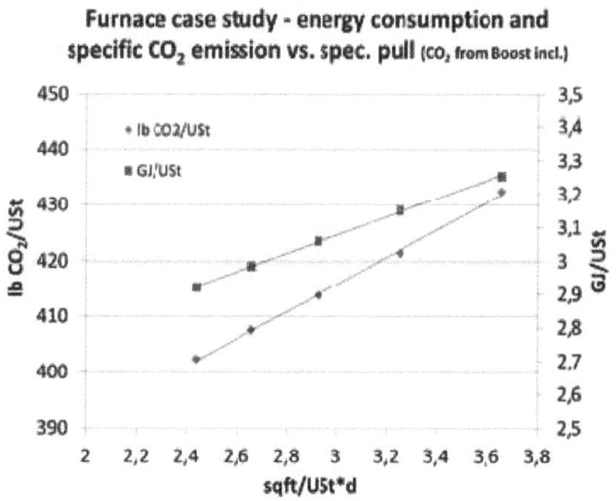

Figure 4. Case Study for Specific CO_2 Emissions and Specific Energy Consumption vs. Spec. Pull

Higher melting performance corresponds with higher earnings and faster return of investment for even higher capital expenditures. The key challenge for SORG is to improve furnace efficiency while lowering specific CO_2 emissions.

NEW FURNACE SOLUTION

The melting capacity of glass melting furnace is dependent on the speed of the solid to liquid transition. The floating batch is a magnitude lower in thermal conductivity compared to the melt underneath. The thinner the batch layer, the faster the heat transfer from hot combustion space and from the melt into the batch piles. High load and thin batch layer requires space inside the melter. Apart from that the more the melt surface is covered with batch the less energy can be transferred from combustion space into the melt. This context is clearly limiting the melting capacity. Electric boosting has typically been applied to compensate for reduced heat flow from combustion space.

Conradt has investigated a larger number of furnaces. He found an empirical relationship between the specific melting performance and the thermal efficiency expressed by the quotation of energy in glass divided by energy supplied to the combustion chamber (Figure 5).

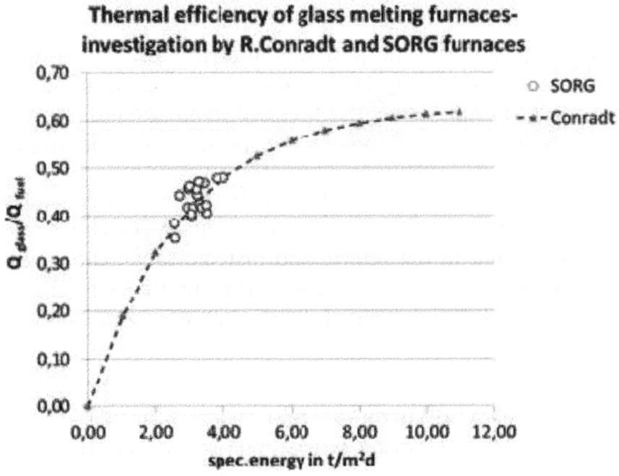

Figure 5. Efficiency Evaluation of Container Glass Furnaces and Our Installation for Comparison (evaluation by R.Conradt, Aachen)

For comparison a number of our furnaces are added in this graph. The conclusion from this is says that very high efficiency requires unrealistic high specific pull. Melt Efficiencies of 3 Gigajoules per ton (3GJ/t) are typically discussed as a goal for the next generation of melting furnaces. 3GJ /t correspond with a thermal efficiency of 0,6 (see Figure 4). It is quite obvious that this will not be achievable with existing furnace designs and only with fossil energy input.

We have investigated different solutions based on CFD modelling. This CFD modelling led to furnace designs in which the charging pocket was significantly enlarged. More space in this section allows an increase of the batch layer without losing free surface in the melter (Figure 6).

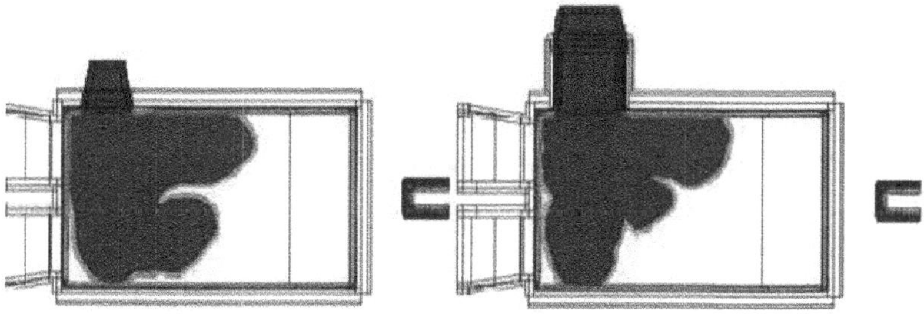

Figure 6. CDF Modelling Results. Batch layer in the standard endport furnace (left) and in the SORG high pull furnace with enlarged doghouse (right)

The larger doghouse in SORG furnace designs has already been proven in many current installations. CFD modelling results are providing valuable information about flow patterns and temperature distribution in this section and suggest that further increase in doghouse size are feasible

The heat losses in the enlarged doghouse design were calculated and due to lower temperatures and air tight connection with the charging machine the heat loss is not higher compared to former open charging pocket conditions.

With the high specific pull conditions the specific energy consumption can be reduced close to 3GJ/t. With flue gas heat recovery the value will be even lower.

In the CFD modelling study about 14% energy input by electric boosting was considered. The specific CO_2 reduction expressed in lb/USt glass can be reduced by more than 20% (Figure 7).

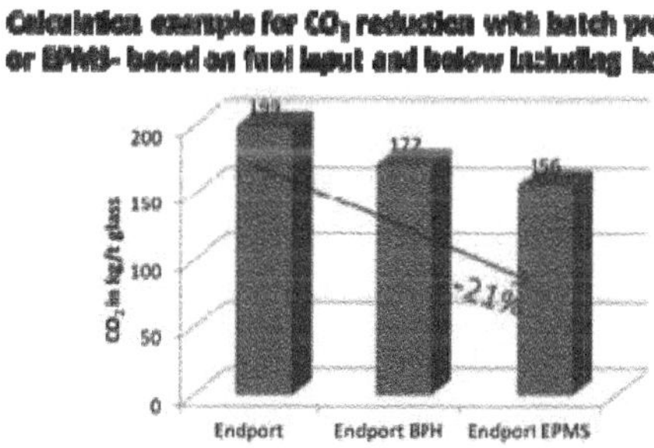

Figure 7. Calculation Example for Specific CO_2- Emissions

Increased Productivity and resultant increased profitability allows further investments for flue gas cleaning systems. The flue gas heat recovery is no longer a disqualifier for applying NOx abatement. Catalytic systems operating at lower temperatures are now available.

SUMMARY

In addition to profitability, environmental issues, legal emission limits and now the CO_2 emissions are the driving forces for ongoing development of glass melting furnace designs. Historically, primary measures like near stoichiometric firing or asymmetrical positioning of the burner nozzles in the port were applied to meet the legal limits for greenhouse gas emissions, resulting in reduction in profitability.

The new challenge for glass manufacturers will be the limitation in CO_2- emissions. Legal limits for greenhouse gases will be further reduced in the near future and will require manufacturers to invest in more efficient catalytic cleaning devices in order to meet emission limits.

The proposed approach to solve these future challenges will be a furnace solution with high specific pull. Special design features will be required to achieve this performance. The significantly increased production rates and resultant increased profitability justifies the additional investment. The high pull on the furnace will also result in a significant reduction in specific CO_2 emissions. Increased use of electric power (i.e. boost) will be a future demand as well since more flexibility in using different energy sources will be required of glass manufacturers.

OXY-FUEL TABLEWARE FURNACE WITH NOVEL OXYGEN- AND NATURAL GAS PREHEATING SYSTEM

T. Görüney, Neşet Arzan
Şişecam R&D, Kocaeli, Turkey
Süleyman Koç
Paşabahçe, İstanbul, Turkey
O. Öztürk and H. Şahin
Paşabahçe Bulgaria EAD, Targovishte, Bulgaria
H. Kim and T. Kang
Air Liquide R&D, Newark, Delaware
Y. Joumani
Air Liquide R&D, les Loges-en-Josas, France
X. Paubel
Air Liquide ALTEC, Paris, France
L. Jarry
Air Liquide, Shanghai, China

ABSTRACT

As the global demand for natural gas continues to rise and environmental regulation targets are becoming increasingly stringent, glass manufacturers are faced with the challenge of reducing costs for melting glass under a variety of operating constraints without compromising glass quality and productivity. This enabled a stronger demand among glass manufacturers to invest in innovative waste heat recovery schemes and energy- and emission reduction technologies.

In order to position well in light of these recent trends, we undertook an effort to implement ALGLASS HeatOx technology offered by Air Liquide to its 200 ton per day oxy-fuel tableware furnace (Furnace B) in Paşabahçe Bulgaria EAD plant, leveraging a co-funding by European Commission LIFE+ program.

Oxy-fuel combustion technology is already proven in the industry to increase production, decrease fuel consumption, and reduce emissions, while maintaining or even increasing product quality; by reducing or eliminating the introduction of inert nitrogen into the furnace. HeatOx technology further extends the aforementioned benefits of oxy-fuel combustion technology by preheating cryogenic grade gaseous oxygen and natural gas via recovery of otherwise-waste heat from flue gas using an intermediate heat transfer fluid, i.e. air.

This paper will share experiences from the field including erection, start-up and operation phases; present most up-to-date energy and emissions performance in 2016, and significant milestones and installation steps achieved towards the implementation of a fully operational HeatOx system, a world first at this scale- and type of glass production.

INTRODUCTION

Şişecam Group operates in flat glass (float), glassware (tableware), glass packaging (container), and chemicals businesses at a total of 44 plants – 19 local and 25 abroad – with over 21,000 employees; nearly half its sales are exports[1]. Şişecam owns and operates a wide variety of waste heat recovery equipment integrated to its air-fuel regenerative float-, container- and tableware furnaces globally. These include waste heat boilers, absorption chillers, steam turbines and organic Rankine cycle systems with a total thermal (heating/cooling) capacity of about 24 MW and net electrical power production of about 17 MW, corresponding to less than 4% of Şişecam's global energy consumption in 2015. While these numbers demonstrate a reasonably well rounded portfolio, there is still room for growth, which constitutes a significant opportunity for further waste heat recovery equipment implementation at Şişecam plants.

Furthermore, there were a number of key drivers for Şişecam to consider extending its air-fuel waste heat recovery portfolio into that of oxy-fuel by applying reactant preheating technology to one of its oxy-fuel tableware furnaces in Bulgaria. Some of these were: bullish energy price outlook at the time of the investment decision[2], increasingly stringent environmental regulations, probability of getting external funding due to the innovative nature of the enabling technology and its application at this scale and type of glass production, and the industrial gas supply synergy thanks to Şişecam's float and tableware production at the same site.

The 2015 United Nations Climate Change Conference, the 21st Conference of the Parties (COP 21) in Paris marked a critical milestone. An important outcome, the Paris Agreement, which will enter into force on November 4, 2016, was signed by 190 states and the European Union as of October 2016. 79 of those parties have ratified or acceded to the Agreement, most notably China, the United States and India, the countries with three of the largest greenhouse gas emissions of the signatories' total (about 42% together)[3].

The main objectives of the LIFE+ Environment Policy and Governance project – Eco-HeatOx – are to reduce specific CO_2- (from combustion alone) and NOx emissions by 23% and 90%, respectively, compared to traditional end-fired regenerative air-fuel combustion, and reduce fuel consumption by 9% compared to oxy-fuel combustion for tableware glass production[4]. Paşabahçe Bulgaria EAD, a subsidiary of Şişecam Group operating in tableware business, is the coordinating beneficiary (project coordinator), while Air Liquide is the associated beneficiary (project partner) of the LIFE Eco-HeatOx project.

The project application was submitted in June 2012. At that time, oxygen supply contract had already been signed and the investment decision to build a new oxy-fuel tableware furnace was already made. The awarded project started in July 2013, the equipment erection started in July 2015, and the system started up in October 2015, followed by a transitional operation phase – mostly debugging and troubleshooting – up until May 2016. Since then, the system has been in steady operation with ongoing improvements made on the fly. The project will end in December 2016 and has met the projected timeline so far.

Reactant preheating technology was described in numerous publications for the tableware application[5-6], and preliminary industrial results from the transitional operation phase were presented by the technology provider[7-8]. This paper's primary purpose, on the other hand, is to share the glassmaker's perspective on the performance and application of this novel oxy-fuel waste heat recovery technology, and share their experience from the erection, start-up and transitional- and steady operation phases.

WASTE HEAT RECOVERY SYSTEM DESCRIPTION

The reactant preheating waste heat recovery system is primarily composed of the reactant heat exchangers, recuperators and burners. The system layout is shown in Figure 1. One key difference compared to the previous float glass references of the said system is that the natural gas- and oxygen heat exchangers were specifically designed to reduce capital cost and footprint.

Specifically, in the float glass reference there were two heat exchangers (one for natural gas and one for oxygen) for each burner. In this installation, however, there are two heat exchangers (one for natural gas and one for oxygen) for each side of the furnace feeding four HeatOx burners.

The multi-channel reactant heat exchangers feed natural gas and oxygen to multiple burners with individual gas temperature control for each burner. The reactant flow and ratio is controlled by flow control skids that handle room temperature natural gas and oxygen upstream of the reactant heat exchangers.

The recuperators are positioned on top of the horizontal flue gas channels on each side of the furnace following the vertical flue gas channels. Due to height limitations and interference with the roof of the building, dual stage recuperators had to be incorporated into the design. There is a duct connecting the flue gas channels of the first- and second stage recuperators. Due to the beam and the floor right above the second stage recuperator, the duct makes two 90° bends before connecting to the second stage recuperator.

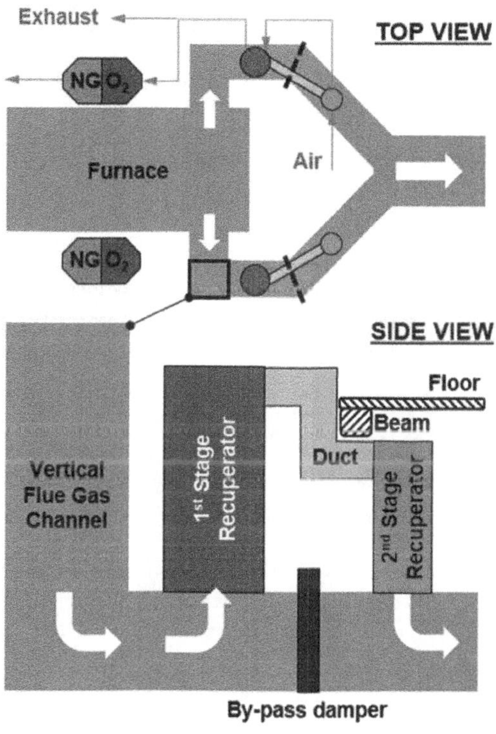

Figure 1. Reactant Preheating Waste Heat Recovery System (HeatOx) Layout

The flue gas can be isolated from the flue gas channel and directed into the recuperators by the by-pass damper positioned in the horizontal flue gas channel between the first- and second stage recuperators. Process air flows in a countercurrent fashion with respect to the flue gas flow and bifurcates prior to the inlet of the reactant heat exchangers to allow hot air to exhaust when the reactant preheating system is not in use. This so-called exhaust mode is also a safety mode in which the recovered waste heat is discharged outside the building in form of hot air in case of an emergency. In this mode, the waste heat is still recovered from the flue gas but it is not used to preheat the reactants.

The HeatOx burners can utilize both hot and cold reactants without requiring a burner change-out, although they are primarily designed to operate at preheated reactant conditions, where the oxygen jet has an optimal velocity and momentum[4]. Burners are of nozzle mix type with oxygen staging capability and separate fuel- and oxygen bodies, fuel being at the bottom, and oxygen at the top. The fuel is distributed, through a refractory burner block, among three injectors arranged horizontally.

ERECTION PHASE

The equipment erection primarily involved the modification of the existing flue gas channels prior to the furnace start-up, relocation of interfering pipework, and installation of heat exchangers, burners and piping for preheated reactants. A remarkable highlight is that the furnace building was originally designed for an end-fired regenerative air-fuel furnace. Therefore, equipment installation required significant construction work and structural modifications under tight spatial constraints in order to comply with architectural limitations, particularly to be able to fit the recuperators. The equipment erection took about three months to complete. Figure 2 shows some instants from the erection phase.

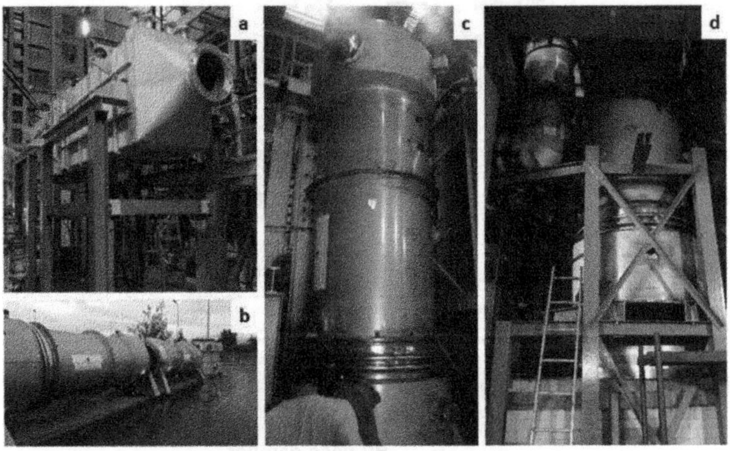

Figure 2. a) Natural Gas- and Oxygen Heat Exchangers b) Recuperator Delivery c) Recuperator Installation d) Recuperators in Place

START-UP AND OPERATION PHASE

System Performance

Based on approximately one year of operating- and troubleshooting experience, no unfavorable effects are observed on furnace pressure management, glass quality and glass color. Same observation applies to furnace aging but it is preferred to wait until the end of the campaign to be able to perform a more accurate assessment on this aspect.

Figure 3 shows natural gas and oxygen preheat temperatures (top two curves) and flow rate (bottom two curves) for R1 burner for a time period of two weeks starting from September 16, 2016. The furnace pull rate during this period was between 176 and 219 metric tons per day. It is remarkable to note that the HeatOx system was adaptable to variations in pull rate and it demonstrated a reasonably steady performance with average preheat temperatures of about 425°C and 475°C for natural gas and oxygen, respectively, which corresponds to approximately 7-8% fuel savings compared to oxy-fuel operation. This includes a 25°C temperature correction to account for the effect of the position of the thermocouple tip on the temperature measurements. The sudden drop of the flow rate near September 20, 2016 is due to a power outage.

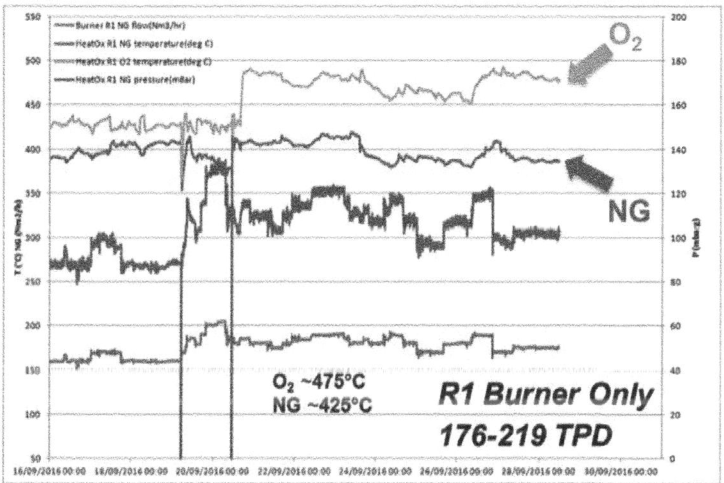

Figure 3. Reactant Preheat Temperature- and Flow Rate Trends for a Two-Week Time Period

Energy and emissions performance of the HeatOx system has been monitored via multiple measurement campaigns. Both flue gas composition and flow rate are measured. Table 1 shows results of three such measurement campaigns. NOx emissions were below 0.3 kg/ton melted glass, which corresponds to more than 90% reduction compared to end-fired regenerative air-fuel furnace operation. These values are well below the emission levels associated with the best available techniques for the oxy-fuel container glass furnaces, which are between 0.5 and 0.8 kg/ton melted glass[9].

Table 1. NOx emissions from various measurement campaigns. NOx expressed as NO_2 corrected at 8% O_2

Date	oct-15	nov-15	may-16
Mode	HeatOx-B	HeatOx-B	HeatOx-B
NOx (ppm)	752.5	760.5	767.1
O2 (%)	2.1	0.65	1.70
NOx expressed as NO2 (mg/Nm3)	1059	993	1057
Pull rate (tpd)	200	200	198
Fumes (Nm3/h)	2113	2113	2133
NOx as NO2 (kg/T)	0.268	0.252	0.273

Troubleshooting

There were several key learnings during the start-up and operation phases of the project that helped Şişecam to better understand the favorable operating regimes, maintenance costs and maintenance schedule of the HeatOx system.

An early discovery during the operation phase was clogging in flue gas ducts. Clogging creates elevated pressure drop and accordingly undesirable loading on the downstream ID fan. Differential pressure measurements were performed followed by an endoscopy campaign to visualize clogging, as shown in Figure 4. SEM/EDS and XRD analyses of deposit samples point primarily to sodium sulfate (Na_2SO_4). The estimated cause of the enhanced clogging is believed to be twofold: i) aerodynamically unfavorable flue gas flow path between the recuperators due to architectural limitations, ii) transitional effects such as low velocity and temperature in flue gas ducts during early operation phase, which are known to enhance deposition through build-up, agglomeration and condensation. To mitigate this issue, design changes are incorporated into the flue gas duct design that enable declogging without interruption to HeatOx operation and with minimal impact to HeatOx performance.

In order to improve the reliability of the system and minimize downtime, the by-pass damper design is revisited on-the-fly to better adapt to process conditions. On-the-fly improvements are made on the flue gas channels to minimize ambient air leaks and reduce thermal wall losses. Hot air leaks out of the process air channels are minimized and thermal expansion issues are fixed.

Carbon build-up was another issue that was observed, most of the time only on one burner position. In addition to creating an extra maintenance load, if unnoticed, it may lead to structural damage on the burner block and the breastwall due to flame deflection, and therefore, it needs to be carefully addressed. En example of this is shown in Figure 5. It is believed that carbon build-up intensifies prior to nonpreheated-to-preheated reactant transition as the fuel injection velocity becomes undesirably high right below the actuation temperature of the reversible and passively acting device that enables natural gas to flow both through inner and outer (annular) natural gas flow passages at and above the actuation temperature. Approaching the actuation temperature from below, an undesirably high velocity slip is created between the natural gas- and adjacent oxygen streams, which causes enhanced recirculation near the natural gas nozzle tips that could favor carbon build-up. At the actuation temperature, however, the natural gas injection velocity decreases suddenly due to a sudden increase in the cross-sectional area of the natural gas flow passages. Below the actuation temperature, the natural gas flows only through the inner nozzle. As part of the ongoing investigation, a nozzle change is under evaluation for this burner position.

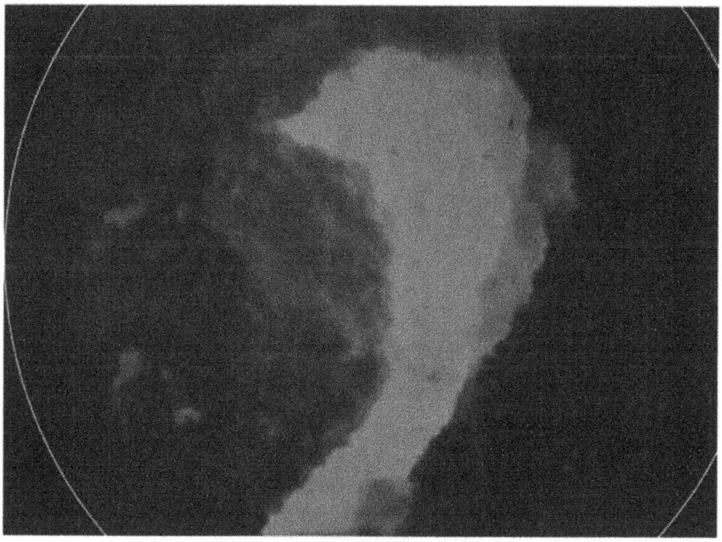

Figure 4. Endoscopy Showing Clogging in Flue Gas Ducts

Figure 5. Carbon Build-up on Natural Gas Nozzle Tip of HeatOx Burner

SUMMARY

A concise summary of the overall system performance for a variety of aspects is given in Table 2 based on the learnings and experience from the project. Additional fuel savings of 1.0-1.5% is expected at the design condition (450° natural gas, 550°C oxygen) with improvements in progress.

Table 2. Summary of overall system performance

Aspect		Experience from Project
Safety	↑	No issue reported
Glass Color & Quality	↑	No issue reported
Furnace Aging	↑	No issue reported so far, need to revisit at end of campaign
Energy (NG and O_2)	↑	- <u>Preheating:</u> ~8%* fuel savings demonstrated on average - <u>Foam:</u> initial results favorable compared to oxy-fuel, further investigation needed for validation
Emissions	↑	Similar to oxy-fuel (>90% NOx reduction wrt air-fuel)
Maintenance / Reliability	○	- Retrofit (architectural constraints, no-WHR mindset) - Learnings (flue gas- and air) → leaks, clogging, dampers - Carbon build-up believed to intensify prior to Cold/Hot transition → ongoing investigation
Ease of Use	↑	Start-up and shut-down with the push of a button Highly flexible and adaptable to variations in process
Project Management	↑	On budget / On time

There were a number key elements and milestones that enabled successful demonstration of this project including but not limited to:

1) External co-funding (LIFE+ 2012) of 50% of the total eligible project budget: The total project budget is 4.3 million EUR, while the industrial WHR prototype cost was only less than half of it, and the LIFE+ program provided 1.7 million EUR funding for this project.
2) Outsourced industrial gas supply synergy: Şişecam has both float- and tableware operations in Targovishte, Bulgaria through its subsidiaries, Trakya Glass Bulgaria EAD and Paşabahçe Bulgaria EAD, which use nitrogen for tin bath and oxygen for melting and polishing, respectively. The combined volume of these industrial gases, which are products of cryogenic air separation process, helped reduce unit cost of supplied gas.
3) Debottlenecking flue gas side issues: These can be summarized as mitigating ambient air leaks into the flue gas channels, providing better thermal insulation on the flue gas channels, introducing more robust by-pass dampers, implementing design changes to enable declogging without interruption to HeatOx operation and with minimal impact to HeatOx performance. It is also advised to avoid transitional regimes to the possible extent.
4) Debottlenecking air side issues: These can be summarized as mitigating process air leaks out of the process air channels and fixing thermal expansion issues.
5) Last but not least, strong commitment and teamwork by all parties involved ensured meeting the projected timeline on budget.

CONCLUSION

World's first preheated oxy-fuel waste heat recovery demonstration at tableware scale is performed at the Paşabahçe plant in Targovishte, Bulgaria. It is validated that the reactant preheating WHR technology does not have any unfavorable effects on glass color and quality, as well as furnace pressure management. Regarding the effect on furnace life, no early signs of accelerated aging have been observed yet, however, it is preferred to wait for the campaign end to be able to perform a more accurate assessment on this aspect.

Approximately 8% fuel savings was demonstrated compared to oxy-fuel, with %1.0-1.5 more savings expected at the design condition (450° natural gas, 550°C oxygen) with improvements in progress. This corresponds to an annual savings of about 250,000 USD to 480,000 USD depending on the natural gas price. More than 90% reduction in NOx emissions was demonstrated compared to the end-fired regenerative air-fuel furnace, at about 0.3 kg/ton melted glass, while the emission levels associated with the best available techniques for the oxy-fuel container glass furnaces is between 0.5 and 0.8 kg/ton melted glass[9].

Şişecam is committed to full compliance with local regulations within its operational domain and in regions where it plans to grow, and positioning in advance of the increasingly stringent environmental regulations by continuing to invest in innovative energy efficiency-, emission abatement-, and waste heat recovery technologies that bring value and reduce environmental footprint of glass manufacturing. Potential NOx regulation change in Turkey in 2017 may also mark an important milestone, if it enters into force, as it is expected to bring along full compliance with the BREF document with specific mass emission limits assigned for different glass segments. This in turn, creates a positive outlook for implementation of oxy-fuel combustion or emission abatement technologies for new investments and retrofits. Şişecam's unique multi-segment production structure at the same site further supports this outlook by creating industrial gas supply synergies that reduce unit costs.

The project also demonstrated the necessity of the technology provider and the glassmaker working closely together, by bringing their own value to the table and leveraging their know-how and expertise, in order to integrate innovative enabling technologies into glass plants. Şişecam will stay committed to reaching energy- and emission reduction targets by leveraging external partnerships in combination with in-house research and technological development capabilities.

ACKNOWLEDGMENT

Şişecam and Air Liquide thank European Commission for supporting Eco-HeatOx project through its LIFE+ funding program. Please visit the www.ecoheatox.com website.

REFERENCES

1. http://www.sisecam.com/en/about-us/sisecam

2. Goruney, T., Huang, R., Wang, J., (2013), "Alternative fuels for glass melting", Glass International, October, pp. 32-34.

3. https://en.wikipedia.org/wiki/Paris_Agreement

4. Kaya, L., Joumani, Y., (2014) "Eco-HeatOx: a novel method to exploit waste heat in oxy/fuel furnaces" 6th Balkan Conference on Glass Science&Technology, Nessebar, Bulgaria, 1-4 October.

5. Kim, H., Kang, T., Kaiser, K., Liedel, S., Jarry, Luc., Paubel, X., Joumani, Y., Kaya, L. (2015), "Heat Oxy-combustion: an innovative energy saving solution for glass industry", 76th Conference on Glass Problems: A Collection of Papers Presented at the 76th Conference on Glass Problems, Greater Columbus Convention Center, Columbus, Ohio, pp. 149-155.

6. Kang, T., Kim, H., Sadasivuni, V., Kaiser, K., Liedel, S., Jarry, L., Leroux, B., Joumani, Y., Kaya, L. (2014), "Oxygen and natural gas preheating at high temperature" 12th ESG Conference, Parma 21-24 September.

7. Jarry, L., Arzan, N., Kang, T., Kim, H., Joumani, Y., (2016), "First industrial results of heat oxy-combustion technology in tableware glass furnace", 24th International Congress on Glass, Shanghai, China, 7-11 April

8. Kang, T., Noel, M., Jarry, L., Arzan, N., Paubel, X., Goruney, T., Orhan, B., Caglayan, E. (2016), "Environmental assessment of reactant preheating technology: Impact on reduction of NOx, CO_2 and dust emissions", 90. Glastechnische Tagung, Goslar, Germany, 6-8 June.

9. Best Available Techniques (BAT) Reference Document for the Manufacture of Glass (2013), p. 340.

ADVANCED HEAT RECOVERY FOR OXY-FUEL FIRED GLASS FURNACES WITH OPTIMELT™ PLUS TECHNOLOGY

S. Laux, U. Iyoha, R. Bell, J. Pedel, A. Francis, K.T. Wu, and H. Kobayashi
Praxair, Inc., Danbury, CT, USA

ABSTRACT

Praxair's OPTIMELT™ Thermo-Chemical Regenerator (TCR) system was first demonstrated on a 50 tpd container glass furnace in September 2014 and has been operating reliably since then. The technology stores waste heat from the hot oxy-fuel flue gas in regenerator beds and uses this energy to reform a mixture of natural gas and recirculated flue gas to hot syngas which is combusted with oxygen in the furnace. Operation of the TCR system on the 50 tpd furnace resulted in 16% to 18% reduction in energy consumption compared to the baseline oxy-fuel furnace, flue gas temperature reduction from 1500°C at the furnace exit to about 650°C after the TCR, and the recovery of about 60% of the sensible heat in the flue gas. To further improve the performance of the system, the TCR technology is being advanced by combining it with another heat recovery step to more efficiently recover waste heat and cool down the flue gas to about 400°C. This advanced heat recovery system, OPTIMELT™ Plus technology, recovers about 75% of the sensible heat in flue gas. This paper summarizes two years of extensive operational experience, performance, and maintenance requirements of the TCR system on the 50 t/d commercial container glass furnace. In addition, the technical concept including engineering layout and results of a numerical model of the OPTIMELT Plus technology is presented for a full scale container furnace.

TECHNOLOGY INTRODUCTION

The OPTIMELT™ thermochemical regenerator (TCR) process is an advanced heat recovery technology for oxy fuel fired glass furnaces. The technology utilizes conventional regenerators and endothermic reforming reactions between fuel and recycled flue gas (RFG) to recover flue gas exhaust heat. For a larger scale commercial furnace, expected fuel savings are about 20% compared to oxy fuel and 30% compared to air-regenerator furnaces. A number of papers [1-5] have already been published to introduce this energy savings technology to the glass industry and report about the commercialization progress. The technology heats and reforms fuel and RFG mixture in a hot checker pack of the regenerator without catalysts or separate steam generation. The syngas created from the reforming of the natural gas contains hydrogen (H_2), carbon monoxide (CO), other hydrocarbons and soot and forms a highly luminous flame which transfers heat efficiently to the glass melt. The reforming process takes advantage of the high H2O and CO2 concentrations (80 to 95% total) in the flue gas of oxy-fuel combustion and synergistically utilizes them as reactants for the endothermic reforming reactions. By comparison, the combined H2O and CO2 concentration of the air combustion flue gas is only about 30% of the flue gas and the remaining 70% N2 cannot participate in the endothermic reactions. The regenerators are similar in design to conventional air heating regenerators, but only 1/3 the size in the checker volume due to the small flue gas volume of oxyfuel combustion, making the retrofit or rebuild an economically attractive option, especially when the available space for a larger furnace is limited. The TCR system is integrated into the oxy-fuel combustion system and the furnaces can be operated either in oxy-fuel firing, or TCR firing mode. The transitions between oxy-fuel and TCR operation are seamless and fully automated in the PLC control.

During normal TCR operation the flue gas from the syngas oxy-fuel combustion is used to preheat a regenerator in heating mode. If TCR is off-line for maintenance or during a power failure, the flue gas leaves the furnace through a separate flue gas port just as it is the case on oxy-fuel furnaces. A key component of this dual TCR/oxy-fuel system is a hot flue gas damper that closes the direct oxy-fuel flue gas outlet of the furnace to divert the flue gas to the TCR regenerators. It closes the opening between furnace and stack when TCR is in service. To preheat the TCR system for service from cold start up this port is gradually closed with the stack damper according to a heating curve that limits the gradient inside the regenerators during preheat. After a desired preheat temperature profile has been established the oxy-fuel burners are shut down while the fuel and RFG flows to the reforming regenerator are ramped up and the resulting syngas is combusted with oxygen in the furnace. The damper for the flue gas is air cooled and designed using CFD and FEA modelling to assure effective cooling and integrity even if power should fail. In this case the damper is automatically opened by gravity to allow reverting the furnace back to oxy-fuel operation with onsite supply of liquid oxygen as a backup. Recent inspection of the damper has confirmed the robustness of the design.

ADVANCED HEAT RECOVERY OPTIONS

The current TCR design recovers about 60% of the sensible heat in the hot flue gas at about 1500°C from the furnace and cools the flue gas to about 650°C. The cooled gas still contains about 40% of the flue gas sensible heat and additional heat recovery is possible by combining TCR with another heat recovery unit. Two examples are shown schematically in Figure 1.

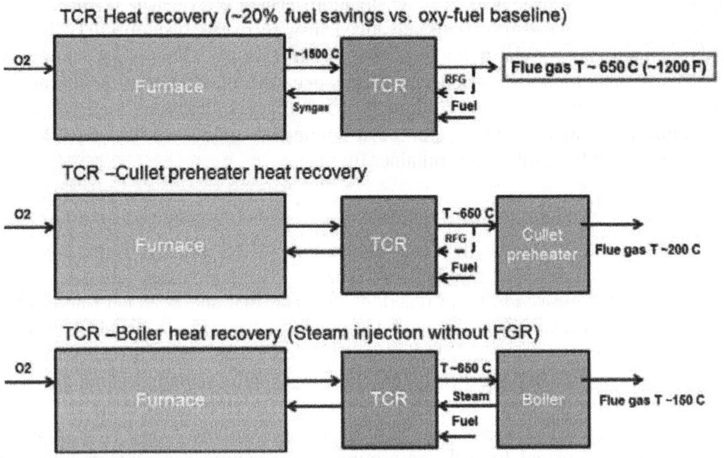

Figure 1. Combined Heat Recovery Options for Oxy-fuel Furnaces

A cullet or batch preheater can be installed to preheat batch/cullet and to cool down the flue gas further from 650°C to a 200°C range. Fuel savings can increase to about 29% relative to the oxy-fuel baseline with this combination. It is also possible to install a waste heat boiler to generate steam and cool the flue gas down to a 150°C range. A portion of the steam can be used as the reforming gas to replace the RFG stream, eliminating the blower-piping system for RFG. The fuel savings remains the same at about 20% in this configuration, but the excess steam can be used for other applications such as power generation. An organic Rankin cycle, rather than steam cycle, can be used for power generation an alternative. Another option is to combine oxygen preheating using a recuperator or regenerators. With a metallic recuperator after the TCR the maximum preheat temperature for oxygen is limited to about 600°C and the combined fuel savings increase only a few percent to 22-23% as compared to the baseline oxy-fuel firing. Regenerative preheating of oxygen [7] in parallel to TCR provides a much higher preheat temperature and enable greater heat recover.

In Table 1 calculated fuel savings for various combinations of heat recovery systems for a generic 300 tpd oxy-fuel fired container glass furnaces are summarized. TCR saves about 20% of the fuel consumption of the baseline oxy-fuel furnace. When oxygen heating regenerators are combined with TCR the fuel savings increases to about 25%. Oxygen preheating regenerators alone reduce fuel consumption by about 11%, which is about half of the 20% fuel savings with TCR alone. The installed costs of TCR and O2 regenerators are about the same since the checker volume is sized based on the total flue gas flow rate from the furnace. Thus, TCR provides more value than O2 regenerators alone.

By combining a cullet preheater with TCR fuel savings as high as 29% are possible. Similarly, by combining a batch/cullet preheater with O2 regenerators fuel savings as high as 28% are possible. However, the value of additional fuel savings have to be compared to the incremental capital investment required to determine the most economic and practical option. Although there are many site specific factors to be considered for any project, our analysis indicates that combined batch/cullet preheating systems are generally most capital intensive. Cullet preheaters are less capital intensive and simpler to operate, especially if clean cullet is available. TCR or O2 regenerators are less capital intensive than the conventional air heating regenerators since the checker volume is about 1/3 of that of the air heating regenerators for the same furnace size. Based on the above cost-benefit considerations and commercial operating and maintenance experience, the combined TCR-O2 regenerators and the combined TCR-cullet preheater systems were selected as most economic options for additional fuel savings.

Table 1. Calculated fuel savings with heat recovery systems for a 300 tpd container glass furnace

Case	Heat Recovery System	Fuel Savings (%)
1	Oxy-fuel	baseline
2	OPTIMELT Thermochemical Regenerator (TCR)	20
3	OPTIMELT Plus (Dual TCR-O2 Regenerator)	25
4	OPTIMELT TCR + Cullet PH	29
5	O2 Regenerator (100% O2 purity, PH Temp 1200°C)	11
6	O2 Regenerator + Cullet PH	22
7	O2 Regenerator + Batch/Cullet PH	28

The dual TCR and O2 regenerator system, termed OPTIMELT™ Plus technology, was chosen for detailed engineering and economic analysis of a 240 tpd oxy-fuel fired container furnace to be rebuilt in 2017. The projected fuel savings compared to oxy-fuel operation of the furnace is 25%, increasing the typical TCR fuel savings of 20% by 5%. A second set of regenerators arranged in parallel to the TCR regenerators is used to accomplish the oxygen preheat to about 2200°F (1200°C). Since the total flue gas flow rate is reduced by the additional fuel savings and the flue gas is split into two chambers, the combined checker volume of the TCR and O2 preheating chambers becomes about 5-10 % smaller, which makes the additional capital cost of this combined heat recovery system relatively small as compared with the baseline TCR system.

A simplified process sketch is shown in Figure 2. Both hot syngas and hot oxygen are then combined in the furnace to form a flame. Oxygen jets are used to give the flame the required direction and length.

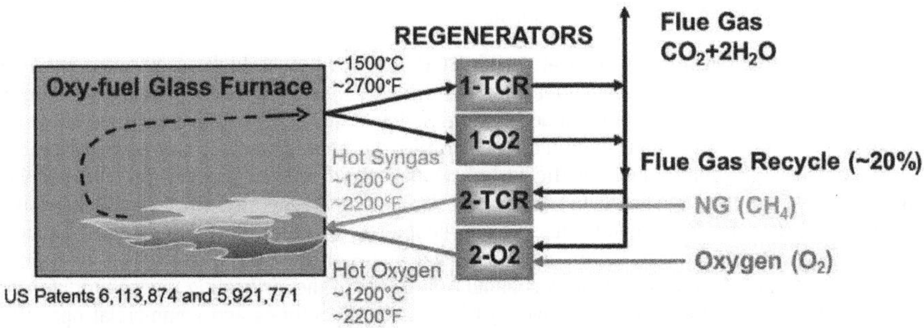

Figure 2. OPTIMELT Plus Regenerative Heat Recovery for Oxy-fuel Furnaces

Table 2 shows the results of an energy balance for a 240 t/d container furnace with 1 MW electric boost and 30% cullet ratio.

Table 2. Example comparison of the two OPTIMELT Heat Recovery Technologies

		Oxy-fuel baseline	OPTIMELT TCR	OPTIMELT Plus
Fuel Input	GJ/t	4.0	3.2	3.0
	MMBtu/ton	3.8	3.0	2.8
Fuel Savings	%	base	21.3	26.6
Flue Gas Temperature	°C	1500	650	400
	°F	2700	1200	750

The additional heat recovery with an OPTIMELT Plus system results in this example in the flue gas being cooled down to a level of approximately 400°C (750°F) versus typically

650°C (1200°F) for TCR technology. This allows reducing the size of the flue gas system due to the lower gas volume.

Simultaneous preheat of fuel and oxidant in regenerators is not new and has been practiced for more than 100 years in producer gas fired glass furnaces. Producer gas is a syngas with low heating value produced from coal gasification with air. It contains approximately 50% nitrogen and when combusted with cold air it results in low flame temperatures not suitable for high volume glass melting. Thus, the gas and air are preheated in pairs of regenerators to elevate combustion temperatures and recover heat from the glass furnace. The regenerators are arranged with two port necks exiting towards the furnace and a sheet-like flame is formed when the gas ignites at the interface between fuel and preheated air. However, forming a flame inside a furnace from the OPTMELT Plus system is a bigger engineering challenge, because hot oxygen and syngas produce an extremely hot flame and a unique burner design is required to avoid port and furnace refractory overheating. An example of CFD simulation for the syngas-hot oxygen flame is presented further below.

UPDATE OF OPERATING RESULTS AT PAVISA

The OPTIMELT TCR technology was installed on Pavisa's Furnace 13 in the summer of 2014 and has been in commercial operation since September 2014. Pavisa manufactures glass and crystal products for global wine, liquor, food, perfume, and pharmaceutical industries. Furnace 13 has a single charger on the left and a nominal pull rate of 50 t/d. This furnace produces a wide variety of glass types and colors with frequent product changeovers involving occasional furnace draining. Cullet ratios can vary widely between 25 to 70% and the furnace pull rate is varied from 40 to about 54 tpd, depending on the type and the color of the glass produced. To understand the efficiency improvement of TCR operation versus the baseline oxy-fuel operation, comparative testing was conducted in 2015 with the same glass type, quality requirements etc. It determined the fuel savings versus oxy-fuel as 18%, which is an expected value for a furnace of this size. Larger furnaces have smaller regenerator wall losses relative to the sensitive heat in the flue gas and more heat can be recovered from the flue gas resulting in a typical TCR fuel savings versus oxy-fuel of 20%. Details of the operation and the very good results with respect to fuel savings, emissions and glass quality are described in [5].

REFRACTORY PERFORMANCE

In addition to the operational performance, the long term performance of the refractory under TCR operation is of specific interest. The refractory in the regenerators is subjected to cyclic changes between reducing and oxidizing atmosphere in reforming and heating mode, respectively. In addition, some experts voiced concerns that silica in the refractory would be reduced by the hydrogen produced during reforming.

Praxair has installed numerous refractory samples in the access ports of the regenerator to collect data on long term performance. Major refractory providers participate in these tests by providing small samples that were imbedded in sample holders at the access doors. Early results indicated that samples installed in the colder (lower) part of the regenerators show no particular differences in performance and it can be concluded that typical refractories that are commonly used in regenerators will work well with the TCR process. However, the various samples in the top of the regenerators above the checkers and in the port neck showed marked differences in performance. Subsequently, refractory testing focused on the samples in the high temperature regions of the process and these tests are continuing in the future.

A second annual inspection of the TCR system at Pavisa was performed in July 2016 and it generally confirmed the findings from the previous year's inspection. The cruciform checkers continue to be in excellent condition. There are no signs of material corrosion. The passages of the checker stack showed no signs of plugging and there was only a slight dusting of the checkers and rider arches in the cold part of the regenerator. The rider arches continue to be in good condition.

However, not all refractory material initially selected for the Furnace 13 installation is suitable for long term TCR operation. The castable material used in the regenerator walls and crown as well as the port necks proved to be an inadequate choice for the conditions due to Nepheline spalling in the high temperature zones. The continued material loss on the upper walls and regenerator crown will eventually result in complete refractory erosion. To prevent extensive downtime it was therefore decided to address this issue before the next cold repair and replace this material in the near future. While the checker passages turned out to be clear from sulfate fouling, the spalling wall and crown refractory results in accumulation on top of the checker pack and a slow increase of regenerator pressure drop. Annual removal of this material mitigates this operational issue and is expected to be resolved with the installation of better suited material in the future.

The flue gas recirculation system that conveys the cooled flue gas to the opposite side regenerator for natural gas reforming or to the stack experiences buildup of typical deposits found in flue gas systems of glass furnaces. These deposits can be easily cleaned during scheduled maintenance without cooling the regenerators down. The material choice for the flue gas piping proved to be adequate and no corrosion is visible after two years.

In general, the condition of the equipment at Pavisa after two years of operation is quite encouraging for the life of the system. The initial concern that cyclic change between reducing and oxidizing conditions in the regenerator could result in premature refractory or checker failure is not supported by our experience. The noted failure of the upper regenerator refractory is due to typical Nepheline refractory failure when material prone to this attack is subject to typical glass batch carryover.

COMMERCIAL OPTIMELT DESIGN FOR A LARGE TABLEWARE FURNACE

The experience at Pavisa is invaluable for the design and construction of larger TCR systems. However, no fundamental changes to the approach demonstrated at Pavisa will be necessary in the future. Praxair is currently working closely with Libbey Holland to install an OPTIMELT system on a large tableware furnace for a 2017 startup [6]. The project was later awarded funding by the European Union under LIFE grant LIFE15 CCM-NL-000121. Currently, the project is in the procurement and fabrication phase and all major components are specified and designed.

The design is very similar to the original Pavisa Furnace 13 arrangement, except that the flue gas system is a forced draft design with baghouse and flue gas desulphurization. The furnace has a side-fired oxy-fuel combustion system and a TCR system in an end-firing arrangement. The oxy-fuel combustion system is a completely redundant back-up to the TCR system. Figure 3 shows the typical end-fired arrangement of furnace, TCR regenerators, flue gas piping and the downcomer with the flue gas damper.

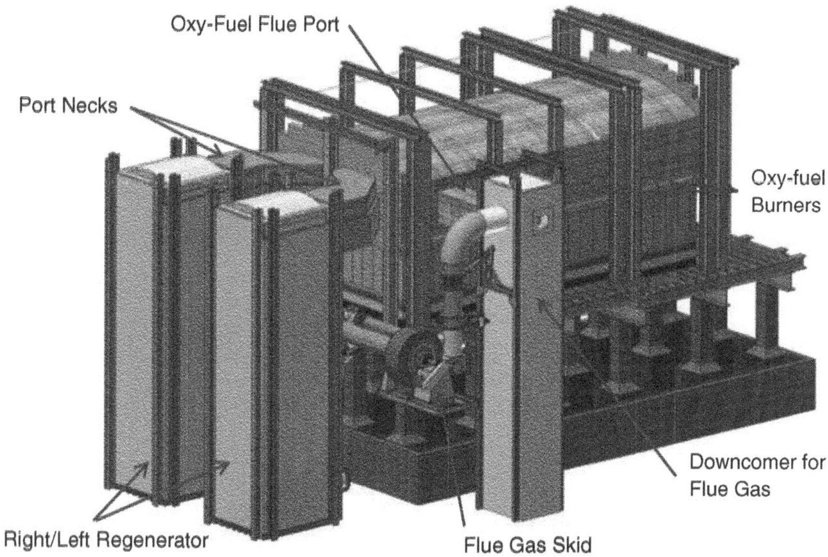

Figure 3. OPTIMELT Arrangement in End-fired Configuration

During OPTIMELT operation, the flue gas leaves the furnace through the regenerator that is in heating mode. A small high temperature fan is used to move the majority of the flue gas to the downcomer where it is diluted with air to a temperature suitable for the downstream flue gas system where a main fan after a baghouse moves the flue gas to the stack. A small amount of flue gas is recirculated after the TCR high temperature fan to the opposite regenerator, where it is mixed with natural gas at the entry of the regenerator. This mixture is then reformed to syngas and directed through the port neck into the furnace where jets below the port provide oxygen for combustion. During this TCR operation the hot flue gas damper inside the downcomer is closed and all flue gas is directed through the TCR system. Should the TCR system be unavailable or experience a shutdown, the flue gas damper between furnace and downcomer is opened and operation can proceed with the oxy-fuel burners. In this case much more air for dilution cooling is required to achieve the desired temperature at the bottom of the downcomer. The dilution air inlets are positioned in the downcomer in a way to manage flow and temperatures preventing damage to the downcomer refractory from the relatively large gas temperature changes between TCR and oxy-fuel operation. The system of air inlets on the downcomer was designed with CFD flow modeling.

The refractory for this project was selected using the experience from the Pavisa technology test and the results from the refractory sample analysis. This led to a differentiated specification of high quality refractory.

OPTIMELT™ PLUS DESIGN CONSIDERATIONS

A typical arrangement of an OPTIMELT Plus system in an end-firing configuration is shown in Figure 4 below. The regenerators for oxygen are located behind the syngas reforming regenerators and separate port necks convey the syngas and the oxygen to the furnace ports in the back wall. Other arrangements such as side by side, all in a row, etc are possible. This specific arrangement was selected to allow space in the center of the back wall for the flue gas stack. The common wall between the reforming regenerator and the oxygen preheating regenerator minimizes the wall heat losses from the regenerators, but requires a few special design considerations to keep oxygen and syngas safely separated until they reach the furnace.

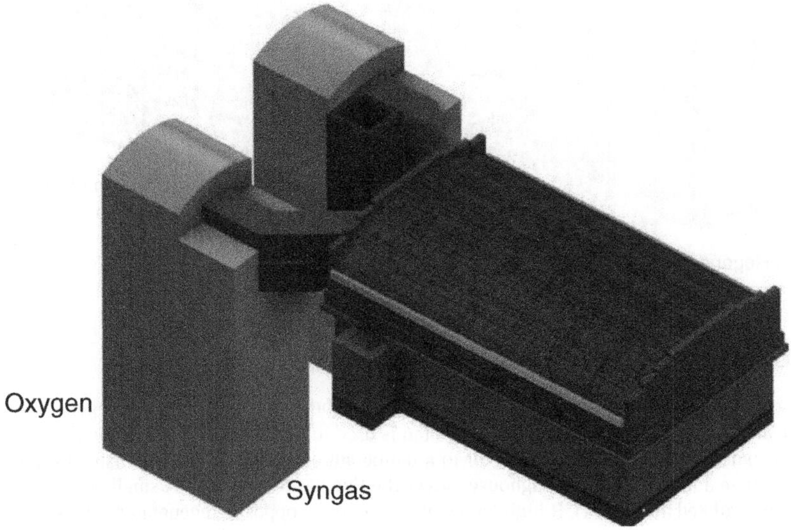

Figure 4. OPTIMELT Plus Arrangement in End-fired Configuration

The time required for designing an advanced combustion system such as the OPTIMELT Plus technology is greatly reduced with the use of computer models. The mixing and combustion of oxygen and syngas in the furnace must lead to a heat transfer to the glass that results in the desired throat temperature at a given pull rate and required glass quality. In addition, the ports, crown or walls should not overheat and flames should not be attached to the walls or lead to large reducing areas on the glass surface. The development of OPTIMELT and OPTIMELT Plus heat recovery technology made heavy use of CFD computer simulation using the GFM software from Glass Service with coupled models for the combustion and the glass space of the furnace.

The hot oxygen and syngas have relatively low velocity when they enter the furnace through the ports. Without any additional momentum these jets would mix right at the furnace rear wall and result in a region of very high temperatures but other parts of the furnace would be too cold. A desirable long flame that extends over most of the furnace length cannot form under these circumstances. Both OPTIMELT technologies makes use of a special burner design feature to produce desirable flame shapes. Our lab tests and the experience from Pavisa 13 confirmed that the unique burner design approach can position the syngas flame at will in the combustion space without the aforementioned undesired effects. The CFD model was not only useful in the design of the combustion system but also for finding initial operating settings which were later optimized during startup by evaluating the visible flames, temperature profiles and the glass quality.

Figures 5 shows CFD model results for the OPTIMELT Plus technology in a 240 t/d furnace. The image shows the temperature contours in the plan view of the furnace cut at the level of the oxygen port. The oxygen enters the furnace at a temperature of approximately 2200°F (1200°C). At the same time the syngas enters the furnace from the separate port. The two gases mix inside the furnace and form a long flame that shows the desired U-shaped profile. The flue gas leaves at the opposite port to preheat the regenerator.

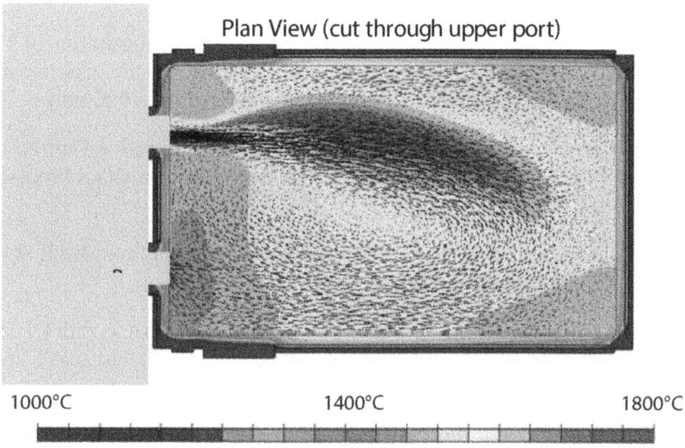

Figure 5. Temperature Contours of Syngas Flame in Furnace

CONCLUSION

Praxair's OPTIMELT™ thermochemical regenerative heat recovery process has been demonstrated in a 50 tpd commercial glass furnace at Pavisa since September, 2014 and the system is in commercial operation. The second annual inspection of the system confirmed the viability of the technology. Checkers and refractory in the cooler parts of the regenerator are showing no signs of corrosion and the checkers stay relatively clean from sulfate deposits. However, the refractory at the top of the regenerator proved to be an inadequate choice and suffers from Nepheline corrosion unrelated to the TCR process. This refractory will be improved in the future.

The next commercial installation of the technology will be on a Libbey tableware furnace. The partial funding of this project by the European Union (reference LIFE15 CCM-NL-000121) is gratefully acknowledged. This furnace will start up in 2017 and provides important momentum for commercial scale-up of the technology towards a broader OPTIMELT application.

Lastly, Praxair is developing enhanced technology, OPTIMELT™ Plus. In addition to recovering energy through reforming of natural gas to syngas this new technology preheats the oxygen to improve energy recovery by an additional 5% to a total of 25% compared with the oxy-fuel furnace, and 35% when compared to a typical air regenerative furnace. The modeling of the combustion process in the furnace shows that the technology can be readily applied to the OPTIMELT Plus technology. The preliminary engineering and costing for a 240 tpd end-port furnace is complete.

REFERENCES

1. Kobayashi, H., Wu, KT, Bell, R.L. (2014), "Thermochemical Regenerator: A High Efficiency Heat Recovery System for Oxy-Fired Glass Furnaces", DGG/AcerS Conference, Aachen, 28 May 2014.

2. Gonzalez A., and Solorzano, E., et. al. (2014), "Commercial Demonstration of ™ Thermo-chemical Heat Recovery for Oxy-Fuel Glass Furnaces", 29th A.T.I.V. Conference and 12th European Society of Glass Conference, 21-24 September 2014, Parma, Italy

3. Gonzalez A., and Solorzano, E., et. al. (2014), "Optimelt Regenerative Thermo-Chemical Heat Recovery for Oxy-Fuel Glass Furnaces", 75th Conference on Glass Problems, 3-6 November 2014, pp. 113-120.

4. Iyoha, U., et. al. (2015), "Thermochemical Regenerator System Proves Itself at Pavisa", Glass International, May, pp. 29-30.

5. Gonzalez A., and Solorzano, E., et. al. (2015), "Operating Experience with OPTIMELT™ Regenrative Thermo-Chemical Heat Recovery for Oxy-fuel Glass Furnaces", 76th Conference on Glass Problems, 2-5 November 2015, pp. 113-120.

6. Praxair Press Release, "Praxair signs contracts to install OPTIMELT™ and oxygen supply systems with leading glassmaker", 12 April 2016

7. US patents 5,921,771 and 6,113,874

IMPROVING ENERGY EFFICIENCY OF GLASS FURNACES

O.Verheijen, M. van Kersbergen, and S. Lessmann
CelSian Glass & Solar B.V.
Eindhoven, The Netherlands

ABSTRACT

Since the beginning of the 20th century major steps have been realized in improving energy efficiency of glass melting furnaces. During this period, energy consumption values for container glass melting decreased from a typical value of about 10 GJ/ton to about 4 GJ/ton (normalized at 50% of cullet) [1]. Although the largest share of potential energy savings have been accomplished, the majority of glass furnaces still operate at energy levels more than 40% higher than what is practically feasible.

To support the glass industry in further reducing energy costs and improving process sustainability, CelSian applies a systematic approach to determine energy efficiency performances of industrial glass furnaces and to quantify the energy savings potential for individual glass furnaces. Different methods and software tools are developed and applied to assess and improve glass furnace energy efficiency. These methods include energy-efficiency benchmarking, industrial validated glass furnace energy balances, and simulation tools to optimize glass furnace design and operation and to improve on regenerator design, and advanced process control technologies.

This paper discusses some components of CelSian's 'energy efficiency improvement platform' with emphasis on energy benchmarking, industrial assessment of glass furnace energy balances and use of in-situ sensors enabling the reduction of energy by means of near-stoichiometric combustion.

ENERGY EFFICIENCY BENCHMARKING

In order to remain a cost-competitive industry, glass companies must continue to focus their efforts on improving glass furnace energy efficiency as one of their key goals. Average primary energy consumption of glass furnaces has decreased considerably until the year 2000, but seems to stabilize since then.However, significant energy savings can still be realized for individual glass furnaces. CelSian applies a systematic approach to determine energy efficiency performance and improvement potential of individual industrial glass furnaces. The first step in this approach is energy efficiency benchmarking [2].

Energy efficiency benchmarking compares the energy-efficiency of (a set of) individual furnaces to the energy efficiency of a large series of glass furnaces present in CelSian's energy-efficiency database. Since 1999, CelSian has performed energy benchmark studies for the international glass industry. On the basis of a limited set of process data (e.g. furnace pull rate, fuel supply, electrical boosting power, cullet percentage, etcetera), the primary energy consumption (normalized to a well-defined cullet-share) for the specific glass furnace(s) is calculated and benchmarked against energy efficiency data of over 420 (anonymous) glass furnaces currently in CelSian's energy-efficiency-database. Figure 1 show the primary energy consumption (normalized to 50% of cullet) for about 230 container glass furnaces ranked from best to worst performance. The best performing container glass furnaces operate at about 3.2

GJ/ton (normalized at 50% cullet), the average actual energy consumption of these furnaces value equals about 4.8 GJ/ton.

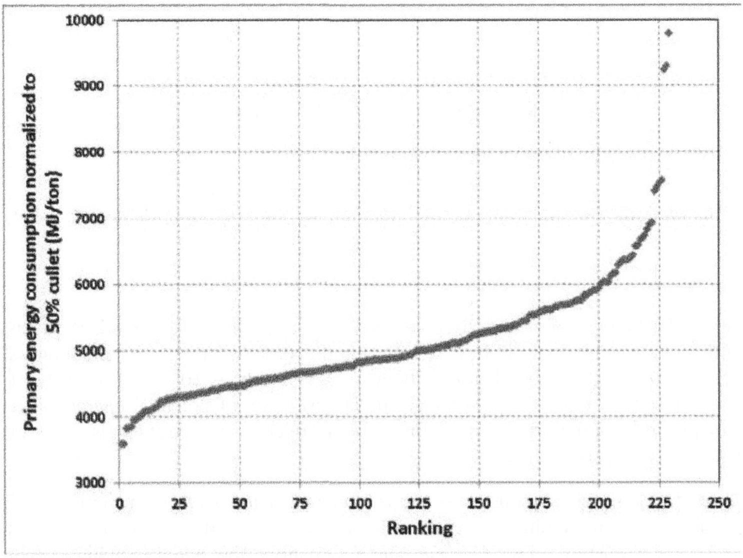

Figure 1. Energy Benchmark Graph Derived from 'Energy Benchmark Database'

During energy benchmarking, for the glass furnace evaluated, the value for the primary energy consumption is calculated and this value is highlighted in the figure shown above. This provides information whether the glass furnace already produces on a low energy consumption level or whether it is likely that a significant amount of energy can still be saved for this specific furnace. Next to this general benchmarking graph, additional graphs can be made, plotting primary energy consumption; actual energy consumption; or energy consumption corrected for cullet share, and/or corrected for furnace age against glass furnace pull rate and specific pull rate. As an example, figure 2 shows the primary energy consumption (normalized) as function of pull rate for container glass. This graph visualizes the trend of lowered energy consumption at increased furnace pull rate.

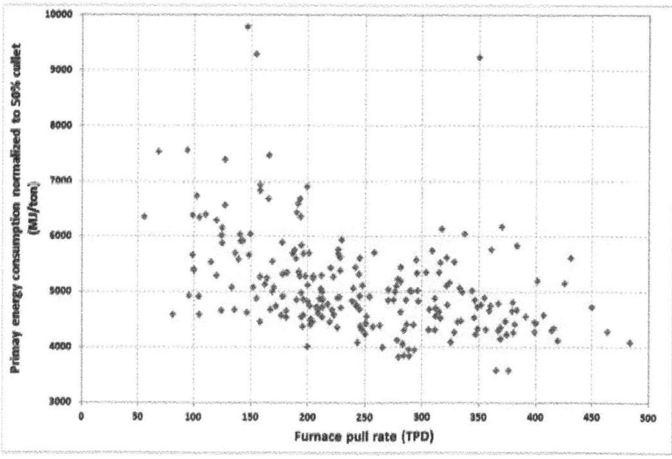

Figure 2. Primary Energy Consumption (normalized) as a Function of Pull Rate for Container Glass (results derived from 'Energy Benchmark Database')

ENERGY SAVINGS POTENTIAL OF INDIVIDUAL GLASS FURNACES

Energy efficiency benchmarking evaluates whether it is likely that energy can be saved for a specific (set of) glass furnace(s). What energy efficiency benchmarking does not do is provide detailed information for where to improve your process to reduce energy consumption. The identification of the most relevant and effective energy saving measures for a specific furnace requires a detailed energy balance. On-site furnace measurements are executed to measure the actual energy performance of the furnace. These measurements include flue gas composition measurements at different positions in the furnace, temperature measurements of flue gas, combustion air, furnace walls, and at other relevant furnace positions. With this information the glass furnace energy balance is calculated distinguishing the amount of energy captured by the batch and glass melt, the amount of energy lost by the flue gases, and the amount of energy lost via the furnace walls and via infiltration air. The structural heat losses are determined for each individual furnace wall with special attention to special items like metal-line cooling, dilatation joint, open holes, and other heat releasing areas. Figure 3 shows an example of the industrial evaluation of heat fluxes through furnace walls and identification of 'weak spots' using an infrared heat measurement camera.

The energy balance derived from the industrial measurements provides a 'copy of the current behavior of the furnace under investigation'. To quantify how much energy can be saved by applying different energy savings measures the use of an energy balance model is required. With detailed energy balance modeling, the theoretical energy balance of a specific glass furnace is set-up using CelSian's detailed Energy Balance Model (EBM) [3]. Based on furnace drawings (including insulation layers and taking into account possible differences in refractory layout between furnace drawings and the actual situation at the glass furnace), actual process settings, and batch compositions, the theoretical energy balance of the furnace is calculated. Generally, the energy balance takes into account the energy flows into and from the melting tank, combustion space and the flue gas heat recovery system (regenerator, recuperator). This theoretical energy balance is validated with the industrial measurements resulting in a calibrated energy balance model that allows 'what-if' scenario's to be performed. Figure 4 shows the user interface of CelSian's 'Energy Balance Model' (EBM).

Figure 3. Industrial Measurements to Determine Heat Fluxes through Furnace Walls and Identification of 'Weak Spots'

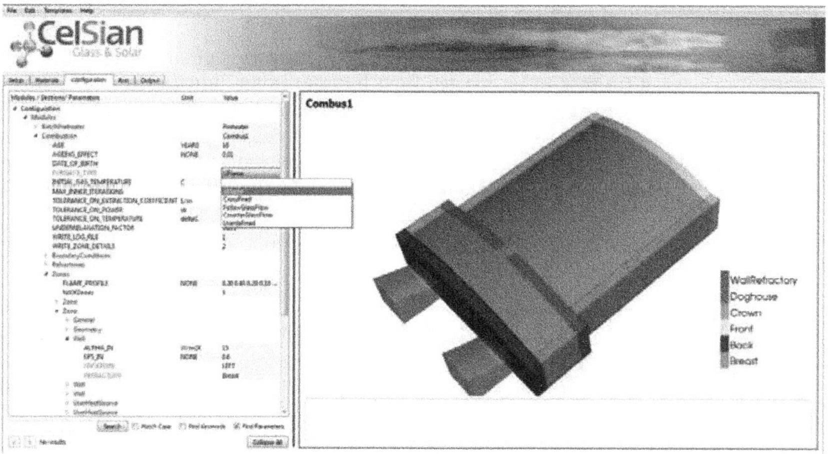

Figure 4. User interface of 'Energy Balance Model' (EBM)

The what-if scenario analysis enables quantification of the potential energy savings for that specific furnace if various boundary conditions or variable are changed, such as: reducing the infiltration air (local); improving the furnace insulation; reducing the excess of air or oxygen; etcetera. To what extent certain measures could be applied is based on the customer and CelSian's industrial experience. Based on long-lasting practical experience with industrial process measurements, CelSian built a database with detailed information about specific local heat losses of furnace walls (side walls, bottom, crown, burner-ports, front walls, bridge walls, etc), regenerator and recuperator efficiencies, overall furnace efficiencies and minimal structural losses. Based on this information, the potential energy savings of the furnace is identified and quantified. The energy losses in a furnace might be a result of:

- Cold air leakages via open joints, openings, or bad sealing due to openings and too low furnace pressures;
- Heat radiation through peepholes, open joints, slits;
- Combustion gas leakages through furnace openings (too high pressure in furnace);
- High excess of combustion air or incomplete combustion;
- High moisture contents of batch and cullet;
- Low cullet fraction;
- Low specific pull rate;
- Flames with low emissivity;
- Low regenerator or recuperator efficiency;
- Bad insulation at different furnace parts;
- Poor glass melt temperature control, resulting in too high average glass melt temperatures in throat;
- Too much cooling water and/or cooling air
- Etcetera.

NEAR STOICHIOMETRIC COMBUSTION

One way of reducing energy consumption is to lower the excess of air (in case of an air-fired furnace) or oxygen (in case of an oxygen-fired furnace) in the furnace. However, while reducing the air/oxygen excess towards near-stoichiometric combustion conditions an increase in CO-concentration in the flue gas is observed. Because of incomplete mixing of the flue gas components, both CO and O_2 exist in the burner port at near-stoichiometric combustion despite the fact that this is thermodynamically unstable. As an example, figure 5 shows the relation of CO-O_2 for both firing cycles in an end-port container glass furnace. As shown by figure 5, the CO-O_2 relation is firing cycle dependent indicating the combustion conditions differ for both cycles.

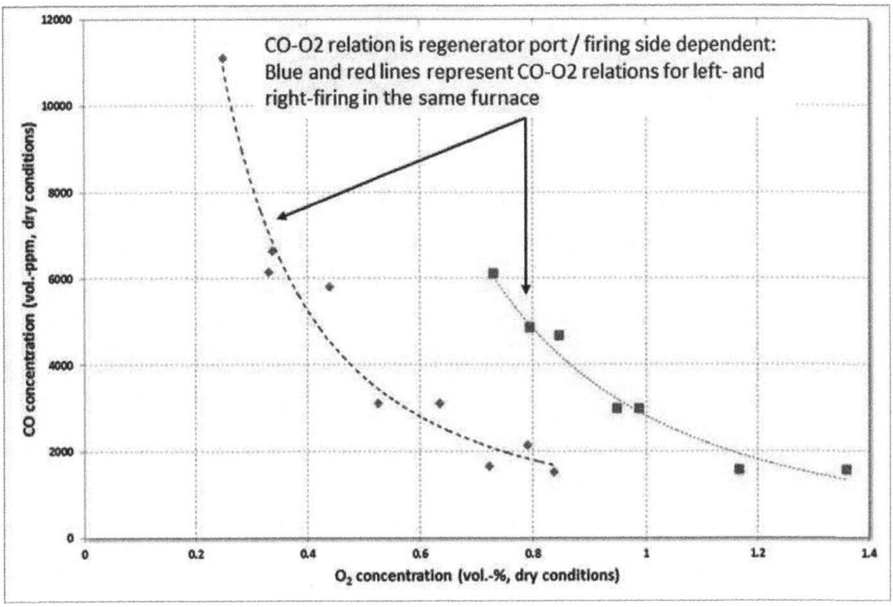

Figure 5. Measured Relation of CO-O_2 for Both Firing Cycles in an End-Port Container Glass Furnace

Figure 6 shows for an end-port fired container glass furnace, the relation between NO_x, dust and SO_x emissions as function of O_2 content in the flue gas. As expected, the NO_x emissions reduce while lowering the O_2-content of the flue gas. However, both SO_x and dust levels increase with lower O_2-content indicating that near-stoichiometric combustion might lead to enhanced evaporation of volatile components that lead to increased crown corrosion and fouling of the regenerator. The latter consequence will directly reduce the energy efficiency of the glass furnace by lowering the thermal performance of the regenerator. Therefore, it is required to monitor (and control) the CO-concentration in the flue gases in case of near-stoichiometric combustion to safeguard furnace lifetime and to ensure improved glass furnace energy efficiency.

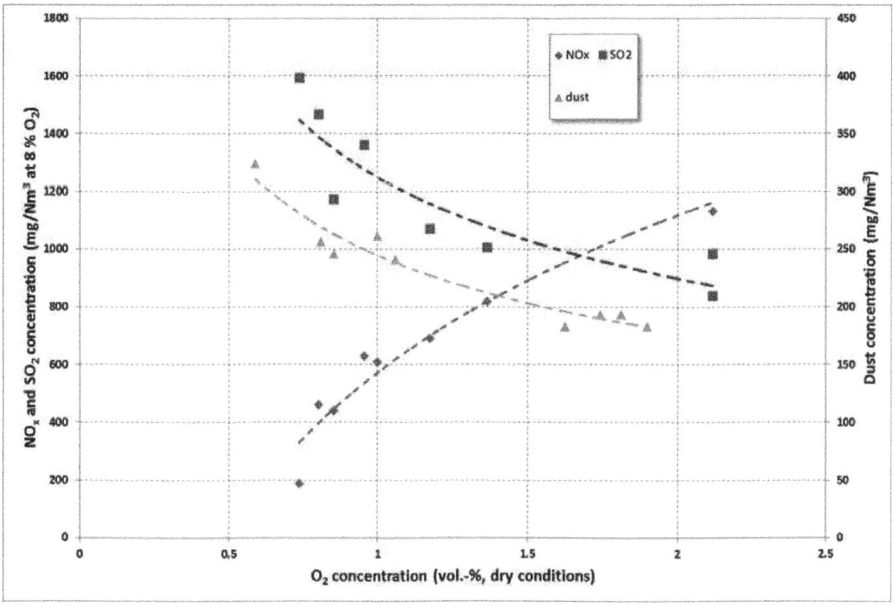

Figure 6. Measured Relation between NO_x, Dust and SO_x Emissions as a Function of O_2 Content in the Flue Gas of an End-Port Fired Container Glass Furnace.

CelSian, in cooperation with NEO (Norway) has developed a new sensor system, the CelSian CO^+ sensor, based on Tunable Diode Laser absorption (TDL) spectroscopy. This new sensor is non-invasive (no direct contact between sensor and flue gases) and provides average, representative values for both the CO- and O_2 concentrations over the cross section of flue gas channels. Moreover, the sensor also measures the water vapor content and the average flue gas temperature, which are critical parameters for mapping the energy flows in glass furnaces.

The newly developed dual laser sensor for glass furnace combustion control has been tested successfully and is now introduced at multiple sites. Industrial tests at regenerative end-port and oxy-fired glass furnaces confirm successful use of this sensor for stabilizing the combustion process. Continuous measurement of both the CO-concentration and the O_2 concentration in burner ports by the CelSian CO^+ sensor provide the extra required input parameters for automatic burner control, in particular at near stoichiometric conditions, constrained by an upper limit CO-concentration.

Figure 7. CO+ sensor installed at oxy-fuel glass furnace

SUMMARY

Since the beginning of the 20th century, even though major steps have been realized in improving energy efficiency of glass melting furnaces, the majority of glass furnaces still operate at energy levels of more than 40% higher than practical feasible. To support the glass industry in further reducing energy costs and improving process sustainability, CelSian applies a systematic approach to determine energy efficiency performances of industrial glass furnaces and to quantify the energy savings potential for individual glass furnaces. This approach is not based on new breakthrough technologies, but aims at improving operational performance by optimized process settings in view of heat recovery, combustion efficiency and minimized structural heat losses. Within CelSian, different methods and software tools are developed and applied to assess and improve glass furnace energy efficiency. These methods include energy-efficiency benchmarking, industrial validated glass furnace energy balances, and simulation tools to optimize glass furnace design and operation and to improve on regenerator design, and advanced process control technologies. The most recent development focuses on the implementation of a newly developed laser based sensor measuring on a continuous bases in-situ both CO and O_2 in burner ports of glass furnaces. This enables automatic combustion control at near-stoichiometric combustion, and thereby saving energy, while maintaining furnace lifetime and regenerator performance.

REFERENCES

[1] R.G.C. Beerkens, 'Trends in Glass Production Innovation or Slowdown', Lecture of the Otto Schott Memorial Medal awardee at the 1st Joint Meeting of DGG – ACerS GOMD, Aachen (Germany), May 25-30, 2014.

[2] R.G.C. Beerkens, J.A.C. van Limpt, 'Energy Efficiency Benchmarking of Glass Furnaces', 62nd Conference on Glass Problems: Ceramic Engineering and Science Proceedings, Volume 23, Issue 1 (2002)

[3] A. Lankhorst, L. Thielen, J. van der Dennen, M. del Hoyo Arroyo, 'Application of an Energy Balance Model for Improving the Energy Efficiency of Glass Melting Furnaces', 74th Conference on Glass Problems: Ceramic Engineering and Science Proceedings, Volume 35, Issue 1 (2014)

OPTIMIZATION OF COMBUSTION SETTINGS: AN ENERGY EFFICIENT APPROACH TO THE REDUCTION OF EMISSIONS BY GLASS MELTING FURNACES

S. Tiozzo, W. Battaglia, R. Dall'Igna, and A. Migatta
Stazione Sperimentale del Vetro Scpa
Murano, Italy

ABSTRACT

Being an extremely energy intensive sector, in the future the glassmaking industry is going to be faced with increasing pressure to lower its polluting (NOx, SOx, acid gases, metals, particulate matter) and greenhouse gases (CO_2) emissions into the atmosphere, and to decrease its specific energy consumption. To achieve this goal, many innovative technological solutions are being developed, such as low-NOx burners, staged combustion systems, new waste gases treatment systems capable of unprecedented efficiency and compactness (e.g. with de-NOx catalyst embedded in the filtering system), or new systems for the advanced recovery of flue gases residual energy. The first step in reaching a better energy efficiency and a lower production of NOx, SOx and CO_2, however, is to simply exploit at its fullest potential the equipment already installed in each furnace, running it in the most optimized way. To support the glass producers in this task, Stazione Sperimentale del Vetro (SSV) has developed an experimental approach for the fine tuning of the burners of glass melting furnaces, leading to sizeable emissions reductions of NOx and SOx, and to slightly improved specific energy consumption performances (and thus decreased CO_2 production).

OPTIMIZATION OF COMBUSTION AT A GLANCE

The optimization of combustion settings of a glass melting furnace comprises several activities, most of which are performed directly on site using a mobile analytical lab, and can be ideally divided into 4 parts:

1) a preliminary collection of furnace background information;
2) a complete characterization of the emissions of the furnace running in the initial combustion setup;
3) the application of a tailored optimization strategy during an in situ experimental campaign, with constant control over the compliance to process and product quality constraints;
4) a new complete characterization of the emissions in the final optimized burners configuration, coupled with a follow up of the mid-long term effects on process temperatures and product quality.

FURNACE BACKGROUND INFORMATION

The first step in the optimization procedure is the acquisition of the furnace "background information", that is a wealth of preliminary data on the furnace design and operation, to obtain a global picture of the current combustion setup and of the process parameters, among which the most relevant are:

- Furnace design (type of melter, burners, heat recovery, dimensions, materials, etc)
- Refractory materials and heat recovery system criticalities (maximum CO level, maximum T, etc)
- Nominal fuel consumption (each burner), Electrical boosting
- Nominal air consumption (each burner), nominal combustion ratio, atomizing air (if any)
- Relevant temperatures (melt, crown, bottom, throat, top regenerator, etc)
- Glass composition and batch formulation, cullet recycling rate, pull rate, color/redox, etc
- Fining agents (evaluation of impact of possible stripping effects)
- Process control system (automatic feedback capabilities)
- Environmental limits for gaseous emissions at the stack

From the in depth review of all this information it is possible to highlight the present criticalities, the degrees of freedom exploitable for the optimization, the constraints and safety/quality requirements to be necessarily met in the final configuration, and thus to start developing a tailored strategy to be applied during the in situ experimental campaign.

INITIAL CHARACETRIZATION OF EMISSIONS

This latter is performed using SSV's highly integrated sampling and analysis mobile laboratories, which are custom built auto-caravans, each equipped with several flue gas analyzers (FT-IR, chemi-luminescence, FID, etc), sampling probes, suction pyrometers, micro-manometers, etc, all connected to a centralized data acquisition and elaboration unit.

For the optimization activities, several parameters of air and flue gases streams have to be continuously and simultaneously monitored, to evaluate in a timely and in depth manner the impacts of the incremental setting changes made to the furnace operation settings.

In particular, sampling probes for gases characterization (T, P and composition) are installed in the following specific positions: for regenerative end port furnaces, inside the port necks, at the top of the regenerator chambers, in the flue gases duct just downstream the regenerator; for recuperative furnaces, at the flue gases inlets and outlets of the metallic recuperators, in the flue gases duct upstream the recuperators; additional discontinuous pressure measurements are performed inside the furnace chamber in both cases, and complete gaseous emissions characterizations are usually performed in the stack before and after the optimization, to assess the achieved improvement level and verify the compliance to emission limits.

At the beginning of the optimization campaign the starting combustion configuration is thoroughly characterized: first, the trends and average values of flue gases composition, temperature and pressure are measured in all the aforementioned critical points; in regenerative furnaces the emission trends are recorded for both right-flame and left-flame conditions during several inversions, and the shape and intensity of "emission spikes" around the flame-switch period are carefully considered.

From all these data several useful information on the furnace behavior and status can already be derived: for example, from the comparison of pressures in the chambers it is possible to identify differences in behavior between right and left side of the furnace/regenerator, while from absolute values it is possible to assess if the clogging/fouling of the checkerworks has reached a critical extent; from the comparison of residual oxygen data it is possible to estimate the position and severity of cold air infiltrations into the system (endoscopic inspection images or thermographic maps can then help in the precise localization of the "leak"); from NOx and CO data the degree of out-of-furnace post-combustion or post-oxidation can be evaluated; from "inversion spikes" analysis it is possible to verify if the flame-switch timing is correct; etc.

After this first global overview of emissions, the furnace characteristic combustion curve, also called "smoke point", is determined: the plot is built by measuring in the port (regenerative) or in the flue gases duct upstream the heat exchanger (recuperative) at each given time the triplet of concentrations of residual O_2, CO and NOx of flue gases, while the combustion settings are changed spanning from over-stoichiometric to almost under-stoichiometric conditions; in particular this can be achieved either by actually changing the combustion ratio, or by changing the furnace pressure level.

Plotting in the same chart for each triplet the CO VS O_2 and NOx VS O_2 points, trends such as the following ones can be obtained:

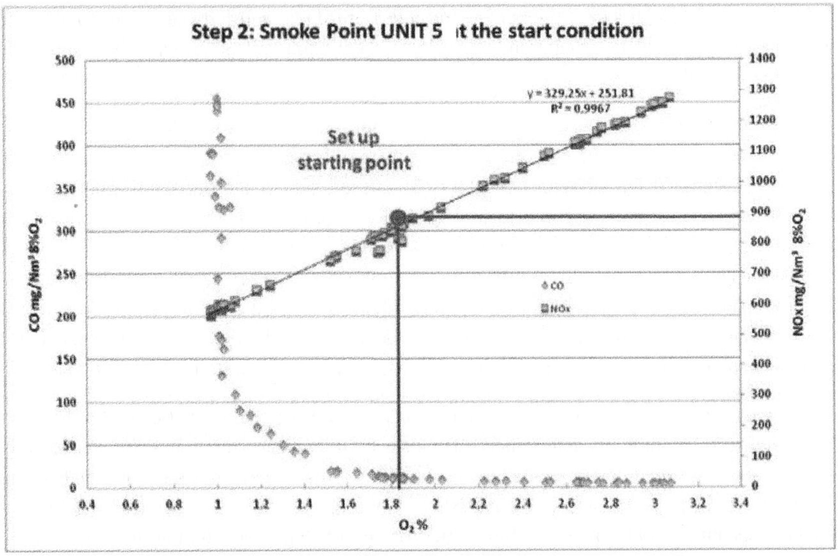

Figure 1. Furnace Characteristic Combustion ("Smoke Point") Curve of a Unit Melter

Smoke point curves are an information-rich tool for the assessment of the behavior of a given furnace setup in case of changes in nominal combustion ratio, since they highlight in a straightforward way what improvements could be achieved by reducing excess air, and which could be the minimum residual oxygen safety level to avoid excessive CO production; moreover, the analysis of curves displacement in case of modification of burners settings (angles, pressures, individual flow rates, etc) is very useful in researching and reaching the optimal configuration.

Once the starting conditions characteristic smoke point curve has been plotted, the so called "burners map" is assessed, that is the individual fuel and combustion air flow rates of each burner are experimentally determined, and any actual deviations from the nominal settings are thus identified. For these evaluations, the furnace overpressure must be set at a suitable level (if it already isn't), in order not to let excessive cold air infiltrations alter the necessary individual measurements.

APPLICATION OF THE TAILORED OPTIMIZATION STRATEGY

From the cross-comparison and analysis of the furnace "background" information and starting combustion setup characterization (i.e. average emissions level and trends, smoke point curve and burners map), a specific optimization strategy, tailored on the system's available degrees of freedom, optimization goals and safety/quality constraints, is finally defined.

The vast majority of settings that are changed and whose impact on NOx emissions and fuel consumption is assessed during a combustion optimization campaign can be grouped in the following categories of "degrees of freedom":

1) Over-stoichiometric air limitation: e.g. reduction of nominal combustion ratio, minimization of cold air infiltrations, increase of furnace pressure.

2) Reduction of combustion rate: e.g. avoid turbulent mixing, delay mixing of air and fuel streams by changing their impingement angle, change flame length and spread, change fuel injection pressure, change burners nozzles.

3) Flame residence time and flue gases recirculation: e.g. asymmetrize combustion to increase flue gases recirculation in the furnace.

4) Minimization of flame temperature: e.g. increase flame emissivity by inducing the formation of soot (delayed combustion, localized under-stoichiometric conditions, injection of small amounts of oil).

5) Tailoring/zoning of combustion density and ratio: e.g. decrease the ratio in burners having high combustion density, work with lower combustion density but higher excess air in burners closer to the waste gases exit from the furnace.

The above degrees of freedom are "explored" by small incremental changes, and their effect is constantly monitored from the mobile lab using the complete analytical system installed on the furnace, especially focusing on sampling points just downstream the furnace (port-neck, flue gases duct before recuperators); when the new average NOx emission level reaches a certain level of improvement, an "intermediate" smoke point curve may be determined, to better assess the behavior of the furnace in the new burners configuration.

The experimental campaign progresses by investigating the impact of each of the available degrees of freedom on emissions (especially composition and temperatures), and once the target level is achieved (usually the compliance with new, stricter emission limits), a complete characterization of the emissions released into the atmosphere by the furnace running in the final setup is performed at the stack.

Instructions on how to maintain an optimized burners configuration even if production needs (e.g. pull increase or decrease) require a change of combustion settings are given to the plant personnel; a periodic furnace audit, including also an endoscopic inspection to check the integrity of furnace and regenerator refractories, is highly recommended to maintain optimization in time.

COMPLIANCE TO PROCESS AND PRODUCT QUALITY CONSTRAINTS

While the aforementioned parameters are experimented and their impact on NOx production and emissions in general is quantified, their compliance to several process, safety and quality constraints is constantly checked, to ensure that no negative impact on production or equipment is produced. For example, some key parameters that are kept under strict control during the optimization campaign are:

- CO level at the exit of the furnace: measured in the port neck and top of the regenerator for regenerative furnaces, in the flue gases duct upstream the recuperators for unit melters; it must be kept at very low values for metallic recuperators (maximum tens of ppm, depending on material) in order not to cause severe corrosion to the heat exchanger; intermediate values can be tolerated by regenerators with magnesite-based checkerworks; higher values (even higher than a thousand ppm) are sustainable for AZS-based regenerator checkers. Every company usually sets a maximum level to be respected, depending on its range of confidence towards high-CO combustion.

- Carryover: measured in the port neck and top of the regenerator chambers for regenerative furnaces, in the flue gases duct upstream the recuperators for unit melters, using an innovative method and dedicated equipment developed by SSV; it can have a severe impact on the heat recovery system and the furnace "outlet" (port, flue gases duct arch) itself, so the new burners set-up should not worsen the amount of particulate stripped from the combustion space, or, where possible, should reduce it.

- SOx emissions: measured in the port neck for regenerative furnaces, in the flue gases duct upstream the recuperators for unit melters; SOx emissions usually tend to increase in case of excessively low combustion ratios, due to the displacement of the sulphates equilibrium by the more reducing atmosphere and higher T; this, coupled with the simultaneous increase in NaOH and Na evaporation from the melt, may worsen the fouling of the heat recovery system (i.e. formation of sodium sulfate deposits gradually clogging the equipment) and require injection of more neutralizing reagent in the acid gases scrubber system, in turn increasing the production of filter dust to be disposed of; the new burners setup should aim at reducing SOx emissions, or, in the worst case, at minimizing their increase (never happened as of today) if lowering other parameters (e.g. NOx) is deemed more important.

- Heat exchanger temperatures: monitored from the furnace control system; in case of unit melter furnaces, the changes in flue gases temperatures produced by the optimization of burners set up must not bring the temperature of the metallic heat exchangers to critical values for the mechanical collapse (creep) or the corrosion of the system; threshold temperatures depend on the material, but are commonly around 830 ÷ 850°C ("background information" on this is mandatory).

FOLLOW UP OF LONG TERM EFFECTS

On a longer term basis, the impact of the new optimized combustion settings on other process parameters must be followed up with the cooperation of the plant personnel (analytical and laboratory support can be provided by SSV):

- Process temperatures: while short term changes in furnace and gaseous streams temperatures can be monitored by the SSV mobile lab, heat recovery systems (especially regenerators) require several days to reach their new equilibrium point, so the final impact on process temperatures and thermal profiles can only be evaluated either by the plant operators (approximated using installed thermocouples) or with a follow-up experimental campaign by SSV (accurate evaluation using suction pyrometers).

- Glass fining: the change in the thermal profile and sulfates equilibrium produced by the change in burners setup may have an impact on the fining reactions in the melt; SSV can provide assistance in the definition of the new optimal mix of fining agents in case blistering problems arise (never happened as of today).

- Glass quality and color: the new combustion conditions must not have an impact neither on glass melt quality nor on glass color or redox state (never happened as of today).

- Energy consumption: fuel consumption by the furnace depends on a huge number of factors, some of which sometimes are not even fully controllable by the furnace manager, but depend on marketing choices, climate, suppliers, etc; therefore, though the nominal energy savings can be easily estimated from the new measured process temperatures, the actual impact on energy consumption can only be assessed on an "average and long term trends" basis.

CASE STUDY: UNIT MELTER FURNACE

In the following pages an example of the results that were achieved during a combustion optimization campaign on a unit melter furnace is reported.

The furnace was a unit melter with a nominal pull of 330 ton/day of flint glass, equipped with two parallel double stage metallic recuperators, with 11 burners per side, burning a high sulfur containing heavy fuel oil. Its furnace characteristic (smoke point) curve in the starting burners configuration has already been reported in Figure 1.

From the analysis of background information, initial emissions characterization and burners map it became evident that some of the burners were deviating from their nominal combustion ratio conditions, that the furnace was infiltrating a lot of cold air from the outside due to a too low melting chamber over-pressure, and that the excess air was disproportionately high, even for the heat exchanger CO-free safety level.

Therefore, the furnace overpressure was increased in small steps, up to a new setpoint minimizing air infiltration and at the same time not impacting downstream processes; the mismatched burners were re-aligned to their nominal ratio; the excess combustion air was reduced in small steps up to a safe CO-free level reducing NOx emissions. The intermediate results on average emissions level are reported in Table 1.

Table 1. Starting and Intermediate average emission levels of the unit melter furnace under optimization

SETUP	Recuperator	O_2 $\%_{vol}$	mg/Nm^3 @ 8% O_2		
			NOx	SO_2	CO
0	Right	2,5	1460	3951	8,3
	Left	2,4	1473	3977	24,4
1	Right	1,94	878	4386	< 1
	Left	2,02	910	4019	< 1

The reduction of excess air, of infiltrations and the re-equilibration of some burners led to a sizeable reduction of NOx production within the furnace, thanks to the decrease in available oxygen (as demonstrated by the 0,5% lower residual O_2 content); the CO content in flue gases at the entrance of the metallic heat exchangers was completely removed, probably thanks to the correction of the settings in the "mis-aligned" burners, that previously created localized strongly under-stoichiometric combustion conditions; SO_2 on the contrary exhibited a slight increase, probably due to the change in sulfates equilibrium brought about by the globally less oxidizing atmosphere (lower average residual oxygen) and by the increase in temperatures of the system (in turn due to the decrease of cold air infiltrations and excess air). The comparison of the smoke point curves determined with the furnace running in these intermediate optimization conditions and in the initial setup is reported in Figure 2.

The NOx VS O_2 curve got lower and less steep, therefore the generation of nitrogen oxides by the furnace became less important and less sensitive to operating conditions; however the O_2 concentration for which the CO generation starts to grow asymptotically shifted to higher values, which means that in the new burners setup the furnace became more sensitive to ratio reductions, and the minimum O_2% for CO-safety increased.

To improve beyond these intermediate results, the second part of the optimization campaign was focused on the strategic tailoring of combustion density and combustion ratio in the various zones of the furnace: burners closer to the throat were set at maximum combustion density (higher temperature) and low ratio, creating local conditions close to stoichiometry (thus low NOx formation due to low excess air); burners close to the flue gases outlet were set at lower combustion density but higher excess air, in order to ensure complete combustion at low temperature (thus low NOx formation) of the CO possibly generated in the other parts of the furnace. The emissions of the furnace running in this new optimized configuration are reported in Table 2, compared with data of previous setups.

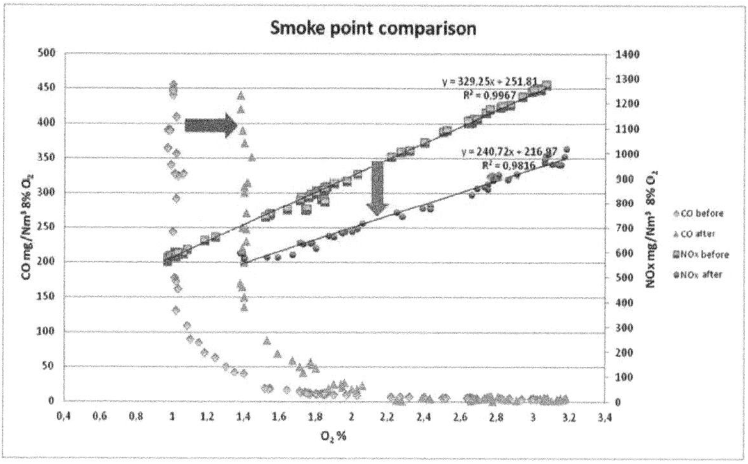

Figure 2. Smoke Point Curve of the Case Study Unit Melter Running in the Intermediate Optimized Setup

Table 2. Comparison of average emissions values in the starting, intermediate and final combustion setup

SETUP	Recuperator	O_2 %$_{vol}$	mg/Nm^3 @ 8% O_2		
			NOx	SO_2	CO
0	Right	2,5	1460	3951	8,3
	Left	2,4	1473	3977	24,4
1	Right	1,94	878	4386	< 1
	Left	2,02	910	4019	< 1
	Bottom center	1,81	894	4220	19
2	Right	1,81	580	3460	< 1
	Left	1,86	598	3267	< 1
	Bottom center	1,37	589	3462	3,5

The strategic distribution of combustion along the furnace axis produced highly beneficial effects on emissions: if compared to the intermediate configuration, the final optimized setup generated -34% NOx and -18% SO_2, with almost complete elimination of CO from flue gases. Thanks to the new combustion conditions, the air preheat temperature increased by approximately 10°C, and the residual oxygen content in the flue gases duct dropped by approximately 0,4%, for a nominal energy savings level that may be estimated in around 0,7% fuel consumption.

The combustion characteristic curve of the furnace running in the final optimized burners configuration is reported in Figure 3, together with the smoke point curves of the previous setups: the NOx trend got even lower and less steep than before, while the asymptotic CO generation moved to much lower residual oxygen content; consequently, the furnace achieved yet more stable combustion conditions, allowing the system to run with utmost safety in globally much lower excess air conditions, with very positive impact on NOx, SOx and CO emission levels. Carryover and particulate matter production were also decreased, while glass quality, evaluated from analyses on follow-up glass products samples, was completely not altered.

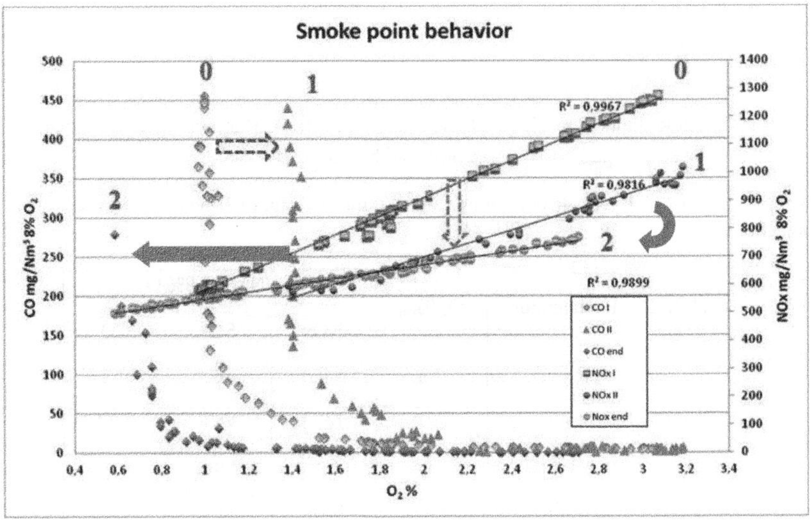

Figure 3. Smoke Point Curve of the Case Study Unit Melter Running in the Final Optimized Setup

CONCLUSION

When faced with the mandatory compliance to ever stricter emission limits, the glass production industry has a wide range of solutions to chose among: various primary and secondary measures have been and are still being developed, each one with its peculiar pro's, con's, efficiency, lifetime and costs. However, the first, cheapest and most straightforward (though not always sufficient by itself) solution is to exploit the installed burners and furnace equipment in the most optimized way, adopting tailored strategies and setups to reduce selected emissions (e.g. NOx) and energy consumption, while at the same time minimizing the impact on the remaining process parameters and product quality.

Melting

IMPACT OF REDOX IN INDUSTRIAL GLASS MELTING AND IMPORTANCE OF REDOX CONTROL

M. Hubert[1], A. J. Faber[1], H. Sesigur[2], F. Akmaz[2], S-R. Kahl[3], E. Alejandro[4], T. Maehara[5]
[1]CelSian Glass & Solar, Eindhoven, The Netherlands
[2]Sisecam, Istanbul, Turkey
[3]Ardagh Glass, Dongen, The Netherlands
[4]Vidrala S.A., Llodio Alava, Spain
[5]Asahi Glass Co. Ltd., Yokohama, Japan

ABSTRACT

The color of a glass is mainly determined by the valence of the multivalent ions it contains. Controlling the oxidation state of the melt is therefore crucial in terms of color stability. The redox state of the melt also greatly influences the fining processes and thus the final quality of the product. In addition, the redox state of the coloring species in the melt affect the radiative heat transfers from the combustion system to the molten glass, and can therefore significantly impact the energy consumption of the furnace. Amongst the most largely produced type of glass, amber glasses owe their specific coloration to the presence of the amber chromophore, a complex between a ferric iron and sulfides. When chromium is added to glasses containing an amber component, colorations such as olive green or antique green can be obtained. Tuning the total chromium content and the redox state are key issues in controlling these colors, and only slight variations may lead to dramatic variations in coloration and to process instabilities, hence to substantial production losses. In this paper, the key parameters governing redox and stability of industrial glasses are presented, with emphasis on amber-containing reduced glasses. Special attention is given to the role of chromium on the color stability. Experimental results obtained in the lab are correlated to theoretical calculations and practical observations in industrial melting tanks. The results and observations presented highlight the crucial role played by redox in industrial glass making, and the importance of its control.

WHAT IS REDOX?

The term redox is often employed in the glass industry, at different places in the process. However, the concept is often not fully understood, and deserves a more detailed explanation. Some elements present in the glass (or melt) composition can be found in different valences, i.e. with different numbers of electrons surrounding them. The most typical multivalent specie found in glass is iron, found mainly as (oxidized) ferric iron Fe^{3+} or (reduced) ferrous iron Fe^{2+}. Other typical multivalent elements in industrial glass melts include sulfur (present mainly as sulfates S^{6+} or sulfides S^{2-}) or chromium (mostly present as Cr^{3+}, but which can be found in reduced form Cr^{2+} or oxidized form Cr^{6+} in specific conditions). The valence of these polyvalent elements in the glass, or oxidation state, depends on several factors, including the concentration of free oxygen available in the melt (often expressed as partial pressure of oxygen, pO_2), glass composition, temperature, and concentration of other polyvalent elements in the glass/melt.

Some examples of redox reactions in glass include (oxidation from left to right, reduction from right to left):

$$Cr^{3+} \leftrightarrow Cr^{6+} + 3\,e^- \quad (1)$$

Or $$Fe^{2+} \leftrightarrow Fe^{3+} + e^- \quad (2)$$

In the glass melt, in presence of dissolved oxygen, these reactions can be written as:

$$Cr_2O_3 + 1.5\ O_2 \leftrightarrow 2\ CrO_3 \quad (3)$$

Or $$2\ FeO + \tfrac{1}{2}\ O_2 \leftrightarrow Fe_2O_3 \quad (4)$$

Redox equilibria depend on temperature and, by consequence, these equilibria evolve during the heating or cooling of the glass melt. In addition, when several multivalent species are present in the melt, they can exchange electrons in redox reactions such as:

$$Cr^{3+} + 3\ Fe^{3+} \leftrightarrow Cr^{6+} + 3\ Fe^{2+} \quad (5)$$

The redox state of a glass (or a melt) is defined as the equilibrium between the different multivalent species it contains, and characterizes the level of oxidation (oxygen concentration level). The amount of dissolved oxygen in the melt depends mainly on the presence and amount of oxidizing species (supplying oxygen to the melt) or reducing species (reacting with oxygen and absorbing it) in the batch.

Specific compounds can be added to influence this redox state. Oxidizing agents such as sulfates, nitrates or polyvalent ions in their most oxidized form, will oxidize the melt by supplying oxygen. Reducing agents such as carbon/cokes, reduced forms of the polyvalent ions, or organic components, will reduce the melt. Indeed, carbon, cokes or organic components react with oxygen to form CO and/or CO_2 gas, which will then escape the melt, reducing its content of dissolved oxygen.

CHARACTERIZING THE REDOX

There are different ways to characterize redox, whether it is expressed as redox state of the melt at high temperature, redox state of the (cold) glass article, or in the batch. In the glass article, the redox state is typically defined as the ratio between Fe^{2+} and Fe^{3+} in the glass, as iron is present in virtually all of glasses industrially produced (added intentionally or present as impurity). It is expressed as Fe^{2+}/Fe_{total} ratio, i.e. the ratio of iron present as reduced iron Fe^{2+} to the total concentration of iron in the glass (wt% Fe_2O_3 in the glass), or as Fe^{2+}/Fe^{3+} ratio. Higher ratio values indicate more reduced glasses, while lower values indicate more oxidized glasses. The concentration of Fe^{2+} can be measured from optical measurements of the characteristic and well-quantified absorption band of ferrous iron in glass samples, centered at 1050 nm in soda-lime-silicate (SLS) glasses [1, 2]. The total concentration of iron in the glass can be measured chemically (e.g. by XRF). The concentration of Fe^{3+} can also be measured optically (absorption band centered at 380 nm), however this measure is not possible in glasses containing an amber component, which masks the characteristic absorption band of Fe^{3+}.

In glass melts, where practical ways to measure optical properties in the melting tanks are lacking, redox is characterized by the $pO_2(T)$, or partial pressure of oxygen in the melt (oxygen partial pressure in equilibrium with dissolved oxygen in the glass melt). Higher pO_2 values indicate more oxidized melts. The pO_2 can be monitored using online sensors typically placed in the feeder channels, and measuring the difference in partial pressure of oxygen compared to a reference [3].

In the batch, the redox state is characterized by a batch redox number (or Simpson redox number) [4, 5]. The contribution of each raw material in the batch (including cullet) is calculated by multiplying its redox factor (negative for reducing species, positive for oxidizing species, 0 for non-redox active species) to the relative weight of this raw material per 2000 Kg of sand in the batch. The sum of the redox contributions from all redox actives species i in the batch gives the final batch redox number, according to:

$$\text{Batch Redox number} = \sum (\text{Redox factor})_i \times (\text{Mass fraction component per 2000 Kg sand})_i \quad (6)$$

The redox factors for selected species are indicated in Table 1 (from [6]).

Table 1: Examples of redox factors for selected reducing and oxidizing raw materials

Reducing components		**Oxidizing components**	
Raw material	*Redox factor*	*Raw material*	*Redox factor*
Cokes (85% C)	-5.70	Sodium sulfate (Na_2SO_4)	+0.67
Fluorspar (CaF_2)	-0.10	Sodium nitrate ($NaNO_3$)	+0.32
Furnace slag	-0.07 to -0.09	Iron oxide (Fe_2O_3)	+0.25
Pyrite (FeS_2)	-1.20	Gypsum ($CaSO_4.2H_2O$)	+0.70

The use of large quantities of external cullet, often done in the container glass industry, makes of redox control a very challenging task, as cullet often contains significant amounts of (reducing) organic contaminations, which nature and amount can vary over time. The higher risk of product color variations with higher cullet content can limit the maximum amount of recycled glass acceptable in the batch. This is especially true for colors such as olive green or dead leaf, which can be produced only in narrow redox ranges.

The redox of the final glass produced does not depend only on the batch redox. The melting conditions in the furnace (atmosphere, residence time, temperature profiles) will also have an influence. There exist however a certain correlation between ranges of values for the batch redox and the glass redox. Specific glass colorations are obtained in typical glass redox ranges, corresponding to typical batch redox values. Industrials producers usually try to maintain or adjust the batch redox to obtain the desired color. Examples of redox values and corresponding colors for soda-lime-silicate glasses are given in Table 2 (from [6].

As compared to the the Fe^{2+}/Fe_{total} ratio measured on the glasses article produced (i.e. after the complete glass making process), measurement of the redox state of the melt inside the furnace allows for an earlier monitoring in the process, enabling faster reaction in case of undesired variations. It must however be noted that continuous measurement of the redox inside the melt at high temperature is challenging and can lead to fast corrosion of the sensors.

Table 2. Typical batch redox and Fe^{2+}/Fe_{total} ranges for different SLS glass colors

Batch Redox	Fe^{2+}/Fe_{total} in the glass	Color
0 ↔ +20	0.10 – 0.40	White / Colorless
-15 ↔ 0	0.40 – 0.60	Green
- 25 ↔ - 15	0.60 – 0.75	Dead Leaf / feuille morte
- 30 ↔ - 20	0.75 – 0.90	Amber

REDOX AND COLOR OF REDUCED GLASSES

The color of a glass is determined by the nature, concentration and the redox state of the (multivalent) coloring ions it contains. The main coloring ions found in industrial glasses include for instance iron, chromium, cobalt, manganese or nickel. Amongst the most largely produced type of glass, amber glasses owe their characteristic coloration to the presence of the amber chromophore, which results from the formation of a complex between a ferric iron and sulfides (Fe^{3+}-S^{2-}) [7, 8, 9]. When chromium is added to glasses containing an amber component, colorations such as *feuille-morte*, olive green or antique green can be obtained.

As the formation of the amber chromophore implies simultaneous presence of both iron in oxidized form (Fe^{3+}) and sulfur in reduced form (S^{2-}), amber-containing colorations can only be obtained in a narrow range of (reduced) redox conditions. This is particularly true for the colors resulting from the combination of amber chromophore and chromium. The amber chromophore forms in reduced iron- and sulfur-containing glasses according to the reaction:

$$S^{4+} + 6\,Fe^{2+} \leftrightarrow S^{2-} + 6Fe^{3+} \qquad (7)$$

This reaction is shifted to the right hand side during cooling of the melt, leading to the formation of the amber chromophore (Fe^{3+} - S^{2-}). Inhomogeneity in the redox state of a glass melt can lead to the formation of color strikes, as illustrated in Figure 1. The glass piece shown presents strikes where the glass is more oxidized, preventing the formation of the amber chromophore (S^{2-} cannot form).

Figure 1. Example of an Amber Glass with Strikes of Different Color due to Redox Inhomogeneity

In the presence of chromium in these reduced melts, additional redox reactions will occur, notably between the iron and chromium. Notably, upon cooling of the melt, Fe^{3+} can be reduced into Fe^{2+} by Cr^{2+}, according to:

$$Fe^{3+} + Cr^{2+} \rightarrow Fe^{2+} + Cr^{3+} \quad (8)$$

Thus, upon occurrence of this reaction (8), the Fe^{3+} present in the melt is reduced into Fe^{2+}, and the formation of the amber chromophore is hindered. Colors such as olive green or *feuille-morte* requiring both the presence of an amber component and a green component brought by Cr^{3+} ions, tuning the chromium content and the redox state of the melt are key issues in controlling these colors, and only slight variations in these parameters may lead to dramatic variations in coloration and process instabilities, hence to substantial production losses in industrial melting tanks.

Reaction (8), explaining the relative instability of the amber component in chromium-containing reduced glasses, supposes the presence of Cr^{2+} in the melts, yet Cr^{2+} is usually not observed in the glasses produced. Thermodynamic calculations have been performed to evaluate the evolution of the redox state of iron and chromium in Fe- and Cr-containing reduced glasses and explain this phenomenon. The results are summarized in Figure 2 (detailed results can be found in [10]). As mentioned previously, redox equilibria depend on temperature. The oxidation state of Fe and Cr ions in the melt at 1500°C are presented at the bottom part of the figure, and the corresponding oxidation state of these ions when the glass in cooled down to room temperature are shown in the top part of the figure. The redox range over which reduced SLS glasses containing an amber component are obtained is also shown. It can be seen that in this redox range, while the glasses at room temperature contain both Fe^{2+} and Fe^{3+}, but chromium only as Cr^{3+}, the corresponding melts at high temperatures will contain also some reduced chromium Cr^{2+}. The results also confirm that when the pO_2 in the melt becomes very low (about 10^{-12} bar at 1500°C), virtually all iron will be present in its divalent form Fe^{2+}, and no ferric iron Fe^{3+} is available anymore to form the amber chromophore.

The calculations also indicate that when pO_2 in the melt reaches 10^{-15} bar at 1500°C, divalent chromium Cr^{2+} can be stabilized in the glass upon cooling to room temperature, and chromium will only be found as Cr^{2+} when pO_2 of the melt reaches values of 10^{-21} bar or lower. These glasses will present a blue coloration, characteristic of divalent chromium. Re-melting such a Cr^{2+}-containing, reduced glass in oxidizing conditions by bubbling with oxygen, will lead to a change in the coloration, as the Cr^{2+} ions will be oxidized into Cr^{3+} ions. The optical properties of such glasses, before and after re-oxidation are showed in Figure 3. The optical features clearly indicate the oxidation of Cr^{2+} into Cr^{3+}, as well as oxidation of Fe^{2+} into Fe^{3+} ions (disappearance of the characteristic absorption of Fe^{2+} ions centered at 1050 nm, and apparition of a narrow absorption band at 380 m characteristic of Fe^{3+} ions). As a consequence, the initial reduced, blue glass turns green after re-oxidation (see [10] for colored pictures of the glass samples). Not only such changes in redox impact the glass color, they can also impact the heat transfers and melting behavior of the glass, as described in the next section.

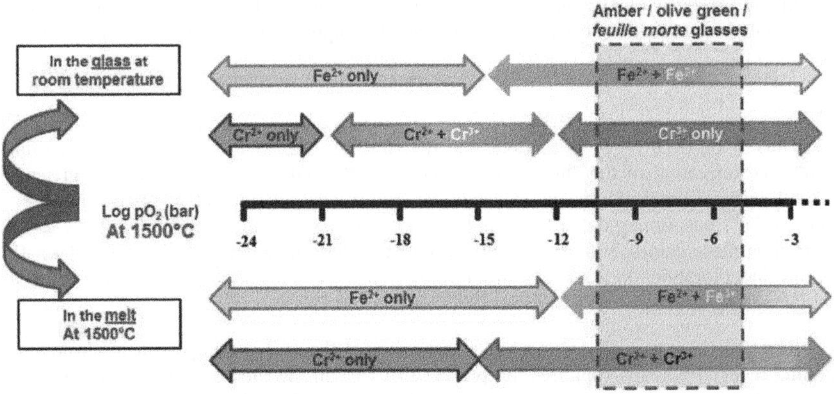

Figure 2. Correlation between pO_2, Redox State and Coloration of Multivalent Ions in Fe- and Cr-Containing SLS Glasses Melted Under Reducing Conditions (from thermodynamic calculations)

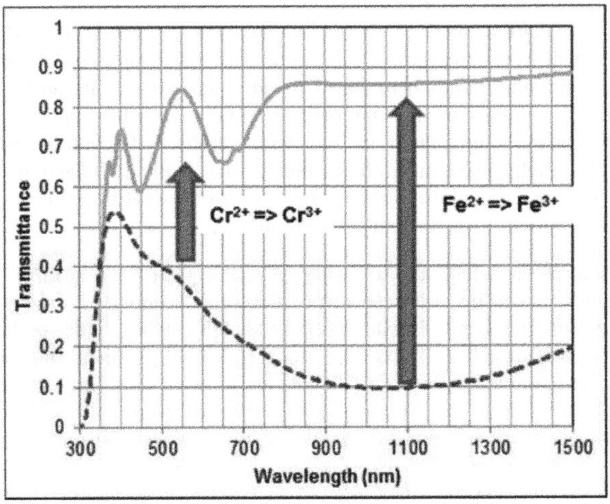

Figure 3. Optical Absorption for a Cr- and Fe-Containing Glass Melted in Very Reduced Conditions (dashed line), and After Re-oxidation by Re-melting with Oxygen Bubbling (plain line). The effect of the oxidation of Fe^{2+} and Cr^{2+} ions into Fe^{3+} and Cr^{3+} ions on the optical absorption bands are illustrated with arrows on the graph.

REDOX AND ENERGY

As mentioned previously, the redox state of a melt will define the ratio of iron in its Fe^{2+} and Fe^{3+} forms, respectively. Besides having an impact on the color of the glass produced, this redox ratio has a tremendous impact on the heat transfers from the combustion space to the glass melt. Indeed, the flames in the combustion chamber and the hot crown of a glass melting tank will emit radiations over a certain wavelength range, mainly in the near infrared (IR). The range will depend on the temperature of the flames and crown. These radiations will then be absorbed by the melt, mainly by the coloring ions it contains. While the exact wavelengths absorbed by a given coloring ion will vary between room temperature and the temperature of the melt (and as the redox state is also a function of temperature), the fact that Fe^{2+} ions absorb significantly more and closer to the near-IR as compared to Fe^{3+} ions remains true.

Figure 4 shows the wavelength range at which black bodies at 1500°C and 2000°C would emit (relative intensities normalized to 1). The optical absorption of iron in a SLS glass melt, for a total concentration of iron of 0.05 wt% Fe_2O_3 and a sample of 10 mm thickness, is shown for the cases where all iron would be oxidized (all iron as Fe^{3+}) and all iron would be reduced (all iron as Fe^{2+}). This graph shows that reduced iron absorbs much more over the wavelength range where the flames and crown emit than oxidized iron. As a consequence, the radiative heat transfers will be different between a more oxidized and a more reduced glass, and a more reduced glass will absorb more heat from the flames. This highlights the importance of redox in terms of heat transfers and energy consumption in glass melting. Indeed, for a glass with a similar composition (fixed iron content), a more reduced melt will have absorbed more heat from the flames.

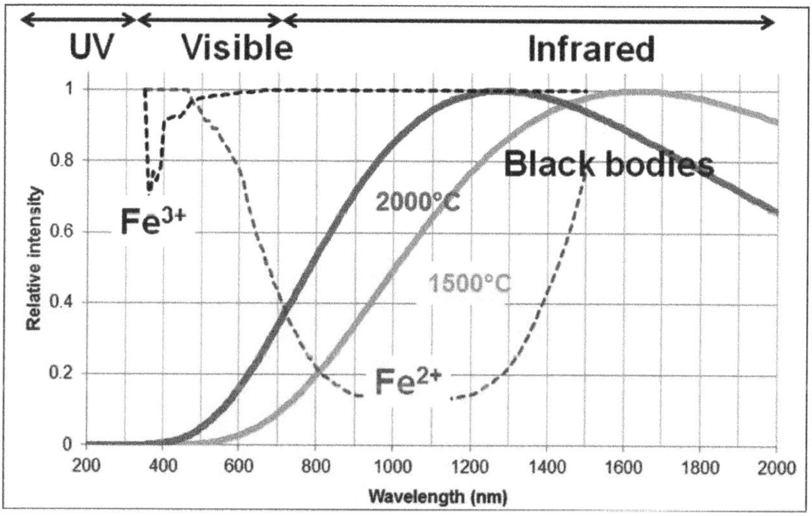

Figure 4. Relative Emission Range for Black Bodies at Different Temperatures and Optical Absorption for a Glass Containing 0.05 wt% Fe_2O_3 either Entirely Oxidized (all ions as Fe^{3+}) or Entirely Reduced (all ions as Fe^{2+})

This could be used to an advantage, for instance by tuning the redox to improve these heat transfers and thus reduce boosting energy needed in the melt. This strategy is for instance applied in some green container glass furnaces, functioning in a more reduced regime to improve energy efficiency. Of course, the color specifications of the product have to be kept in mind, and the redox controlled in order to ensure simultaneously optimized energy efficiency and good product quality.

The impact of the melt redox state can also be modeled. CFD modeling studies have been performed for a container glass furnace producing an emerald green glass, containing 0.25 wt% Fe_2O_3 and 0.30 wt% Cr_2O_3. Two cases have been modeled: an oxidized case, with a Fe^{2+}/Fe_{total} ratio of 20%, and a reduced case with a Fe^{2+}/Fe_{total} ratio of 60%. This Fe^{2+}/Fe_{total} ratio corresponds to the redox state of the melt at the level of the batch feeder (e.g. if the batch was composed of 100% cullet with this redox state). The results of the modeling of the thermal conductivity in the melt for both cases are shown in Figure 5. The reduced melt, containing significantly more Fe^{2+} (for a similar total iron content) than the oxidized case, shows a significantly lower thermal conductivity. In other words, the reduced melt absorbs much more radiation, especially in the top part of the melt, as compared to the oxidized melt.

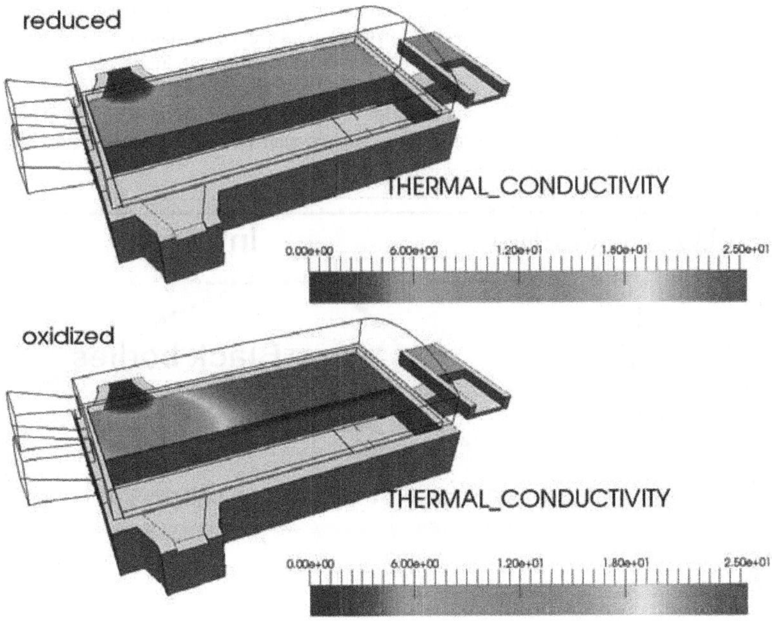

Figure 5. Simulated Thermal Conductivity in an Emerald Green Container Glass Melt, either Reduced (Fe^{2+}/Fe_{total} Ratio of 60%) or Oxidized (Fe^{2+}/Fe_{total} Ratio of 20%)

As a consequence of this difference in thermal conductivity, and keeping similar combustion settings in the furnace, the calculated bottom and throat temperatures are 12°C lower in the case of the reduced melt (graph not shown in this paper). This example illustrates the potential consequences of a change in the redox of the melt: such a redox variation will impact the heat exchanges between the combustion space and the melt, leading to variations in the temperature profiles and potential variation on the energy input required for maintaining stable production.

On the other hand, this also illustrates that redox can be tuned, to a certain extent and depending on the glass being produced, to optimize heat transfers and save on the energy required for the melting. This type of strategy is successfully implemented in some plants, with a redox optimization leading to savings on the boosting energy required to melt the glass, as reported by Ardagh their emerald green glass (85% cullet in the batch) melted in more reduced conditions to favor better heat exchange between the flames and the melt [11]. The color specifications required for the product, also dependent on the glass redox, have however to be kept in mind, and there can be limitations to the extent to which redox can be tuned for optimizing heat transfers. In addition, redox has a strong impact on the fining behavior of the glass melts, as described in the next section.

Of course, any intentional modification of the redox state of the melt needs to be correlated to the specifications of the glass produced, as redox state also influences the product color. When moving to more reduced glass melts, the higher is the possibility to lead to the formation of amber streaks in flints and green glasses. Therefore, the closer from the acceptable redox limit in terms of coloration of the products, the tighter should be the control of the redox of the melt. In that sense, the use of inline redox sensors, placed in the feeders, can be of high practical value.

REDOX AND FINING

Redox has a major impact on the sulfate chemistry, and thus on the mechanisms of fining in sulfate-fined glasses (which represent over 90% of the glasses produced in the world) [12]. Sulfur can be present in the glass melt in different oxidation states: as sulfates (S^{6+}, or SO_4^{2-}) in more oxidizing conditions, as sulfides (S^{2-}) in more reducing conditions, and possibly as sulfite (S^{4+}, or SO_3^{2-}) in an intermediate redox range. The solubility of sulfur in the glass depends on redox, following a trend described by Budd in the so-called Budd curve (see Figure 6, data from [9]).

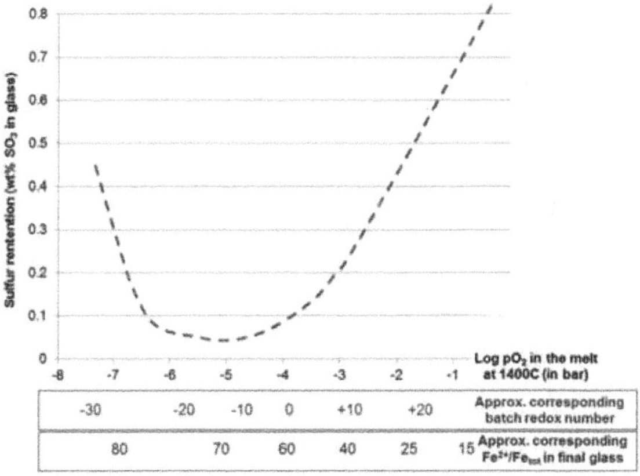

Figure 6. Example of a Budd Curve for a Soda-Lime-Silicate Glass

Sodium sulfate (Na_2SO_4), or salt cake, is the most widely used fining agent and source of sulfur in the glass industry. Sulfur can also be present as impurity in some raw materials, and contribute to the fining. A significant amount of sulfur may also come from the cullet used in the batch. In this case, the amount of sulfur depends not only on the amount of cullet used, but also on the cullet type and redox state. Indeed, reduced glass cullet (such as amber glass cullet) will contain typically around 0.03 wt% SO_3, while oxidized flint cullet may contain more than 0.15 – 0.20 wt% SO_3.

The fining agent will release gases at high temperatures (mainly SO_2 and O_2), when their partial pressure in the melt exceeds 1 bar. The temperature at which the sulfate fining mechanisms activate (the fining onset temperature) is highly dependent on the redox state of the melt. Using CelSian's Gibbs tool, the fining onset temperatures for the oxidized and the reduced emerald green container glasses described above can be calculated. The results of the calculations are shown in Figures 7 and 8 for the oxidized and the reduced melt, respectively. It can be seen that the calculated onset temperature (temperature at which the partial pressure of SO_2 and O_2 reach 1 bar or 10^5 Pa) is decreased by more than 200°C for the reduced melt (1121°C vs. 1329°C for the oxidized melt).

A variation in the redox state of the melt will thus greatly influence the fining behavior of the melt, leading to episodes of insufficient or excessive fining and formation of seeds and blisters in the final product (in addition to possible variations in the glass coloration and in the heat transfer profiles). Often, there exist an optimum redox range for a given melt in which the seed/bubble count is the lowest, and sudden increase in seed/bubble count is observed when melting in more reduced or more oxidized conditions, highlighting once again the importance of a good redox control in glass production. This is especially critical for reduced glasses containing an amber component such as olive green or antique green glasses [13].

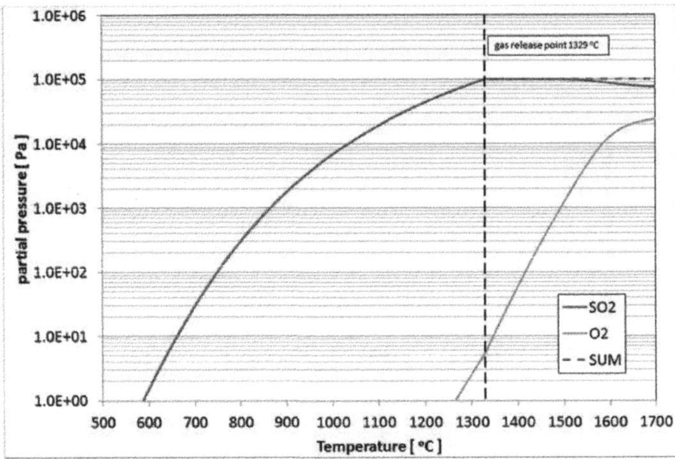

Figure 7. Calculated Fining Onset Temperature for the Oxidized Glass (Fe^{2+}/Fe_{total} Ratio of 20%)

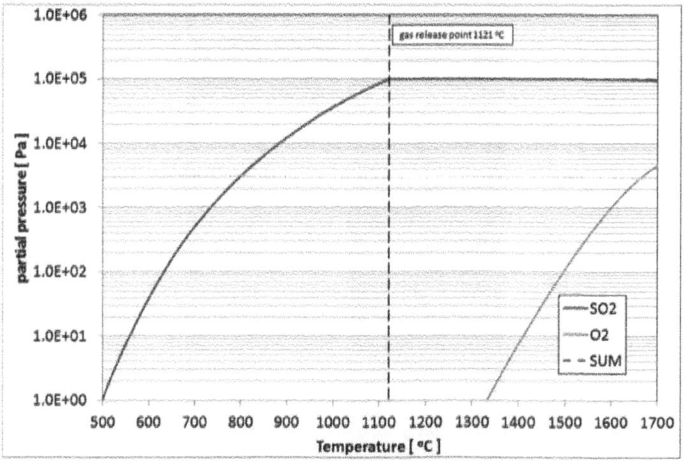

Figure 8. Calculated Fining Onset Temperature for the Reduced Glass (Fe^{2+}/Fe_{total} Ratio of 60%)

In addition, the change in the fining behavior of the melt with a variation in redox may lead to foaming on top of the melt [14, 15]. The foam will act as an insulation layer, hindering the heat transfers from the flames to the melt, and leading to increased energy consumption and reduced bottom temperatures. Therefore, this aspect also has to be kept in mind when optimizing the redox of a melt, as the expected benefits from melting in more reduced conditions in terms of heat transferred can be balanced off by such foaming behavior.

The fining and foaming behavior of glass melts with different redox states can be studied experimentally using a High Temperature Melting Observation System – Evolved Gas Analysis (HTMOS-EGA) [16]. Figure 9 shows two experiments conducted on glasses with similar batch composition but with different redox states. At high temperature, a clear difference in the foaming behavior of both batches can clearly be seen.

Figure 9. HTMOS Observation of Two Glass Melts with Similar Composition but Melted at Different Redox States. The pictures are taken at the same temperature in both cases

CONCLUSION

Redox plays an essential role in different steps of the glass-making process. Indeed, it has a direct effect on the energy transfers in the melting tank, on the fining chemistry in the melt, and on the final color of the article produced. It is directly affected by the composition of the batch, included both the oxidizing or reducing components intentionally added and the potential organic contamination it contains, mainly from external cullet. The control of the redox therefore appears as critical at the different stages of glass melting.

There is not one single type of reaction a furnace will show to a change in redox (may it be planned or may it come from an unexpected contamination event). It therefore appears essential to have a good knowledge of the behavior of a given furnace to a redox variation, in order to select the appropriate actions to take, or to optimize the furnace operation. Notably, optimizing the redox of the glass can be used (to a certain extent, and depending on the type of glass produced).

Several strategies have been developed in order to characterize redox of the batch, melt and glass. The earlier the redox is controlled during the process, the faster a response can be implemented. The redox state of the batch, which can be defined using the so-called Simpson batch redox number, is the earliest mean of redox control in the industrial process. However, in case a large amount of recycled cullet (likely to be contaminated with varying amounts of organic compounds) is used in the batch, the actual redox state of the batch may fluctuate, making a precise control very complex in such cases (e.g. for container glass production).

In parallel to the strategies, CFD simulation tools such as CelSian GTM-X or Gibbs' tool can be used to model the behavior or a furnace and the impact of redox. The HTMOS-EGA setup also allows for experimental evaluation of the impact of redox on the fining and foaming behavior of a given glass melt. Each melt and furnace have their optimum redox conditions. A good knowledge of the impact of redox on the different parameters redox affects is critical, and a tight redox control essential in order to ensure optimum industrial glass production.

REFERENCES

1. A. Weyl, Coloured Glasses. 1992, Sheffield, England: Society of Glass Technology.

2. C.R. Bamford, Colour generation and control in glass. Glass Science and Technology. Vol. 2. 1977, Amsterdam: Elsevier Scientific Publishing Company.

3. P. Laimbock, In-line oxygen sensors for the glass melt and the tin bath, GlassTrend – ICG seminar on Innovation in glass production, April 2013, Eindhoven (Netherlands)

4. W. Simpson and D.D. Myers, The redox number concept and its use by the glass technologist. Glass Technology, 1978. 19(4): p. 82.

5. M. Nix and P. Williams, calculation of the redox number of glass batches containing recycled cullet. Glastechnische Berichte, 1990. 63K(271-279).

6. R. Falcone, S. Ceola, A. Daneo, and S. Maurina, The Role of Sulfur Compounds in Coloring and Melting Kinetics of Industrial Glass. Reviews in Mineralogy and Geochemistry, 2011. 73(1): p. 113-141.

7. P. C. Ross and D. D. Myers, Amber Glass - 40 Years of Lessons Learned, in A Collection of Papers Presented at the 66th Conference on Glass Problems: Ceramic Engineering and Science Proceedings. 2008, John Wiley & Sons, Inc. p. 129-139.

8. R. Beerkens, Sulphur chemistry and sulphate fining and foaming of glass melts. Glass Technology, 2007. 48(1): p. 41-52.

9. R. G. C. Beerkens and K. Kahl, Chemistry of sulphur in soda-lime-silica glass melts. Physics and Chemistry of Glasses, 2002. 43(4): p. 189-198.

10. M. Hubert, A.J. Faber, F. Akmaz, H. Sesigur, E. Alejandro, T. Maheara, and S.R. Kahl, Stabilization of divalent chromium Cr(II) in soda-lime-silicate glasses. Journal of Non-Crystalline Solids, 2014. 403: p. 23-29.

11. S. Kahl, Energy efficiency improvement by better redox control, GlassTrend – TC09 seminar on furnace heating technologies and energy efficiency, April 2015, Bilbao (Spain)

12. R. G. C. Beerkens, Sulfate Decomposition and Sodium Oxide Activity in Soda–Lime–Silica Glass Melts. Journal of the American Ceramic Society, 2003. 86(11): p. 1893-1899.

13. A. Alejandro and A. Serrano, An in-line redox sensor performance test on reduced glass redox control, 90th DGG Tagung – GlassTrend meeting, June 2016, Goslar (Germany)

14. P. Laimbock, Foaming of Glass Melts. 1998, PhD Thesis at the Technical University of Eindhoven.

15. A.-J. Faber, O. Verheijen, and J. Simon, Redox and Foaming Behavior of E-Glass Melts. Proceedings of the 7th International conference on advances in fusion and processing of glass 2004. 14: p. 71-82.

16. M. Rongen, M. Hubert, P. Marson, S. Lessmann, O. Verheijen, and R. Bei, Laboratory facilities for simulation of essential process steps in industrial glass furnaces, 75th Conference on Glass Problems: A Collection of Papers Presented at the 75th GPC, 2015.

Modeling and Forming

FLUID DYNAMICS ANALYSIS LEADING TO INNOVATIVE GLASS HOMOGENIZATION DEVICE "ARCTWISTER"

A. Fuchs
Umicore AG & Co. KG, Platinum Engineered Materials
Hanau, Germany

ABSTRACT

The quality of the glass melt is determined by thermal and compositional homogeneity across the feeding channel. Achieving that homogeneity with stirrers made from precious metals as well as from refractory is a widespread method used in the glass industry. Homogenization of glass melts by stirring happens under laminar flow conditions due to the low Reynolds-numbers at typical stirring temperatures and viscosities. As the glass viscosity conditions do not permit turbulence, good homogenization is hard to achieve. Compositional and thermal differences such as cords cannot be fully dispersed, they can only be stretched and folded until they are virtually invisible. This poses the question: What stirrer geometry is fluid-dynamically efficient as well as considering a multitude of other requirements such as low precious metal weight, longevity and robustness. In a project carried out by Umicore Platinum Engineered Materials, the flow inside a stirring cell was studied by computational fluid dynamics and predominantly by physical simulation in order to develop an advanced stirring device. Cord was simulated by colored tracer liquid at different locations in a glucose model and different stirring concepts were tested and compared. The exercise revealed valuable insight into how the glass flow should be affected and what geometry an efficient stirrer should have when it can be made from oxide dispersion strengthened PtRh-alloy. Finally, the study led to the development of the ARC*twister*, a modular, self-supporting stirrer made from fine grain stabilized FKS®PtRh-alloy. The stirrer creates a unique S-curved flow characteristic moving the glass volume vertically. It also transports the glass horizontally from the outside to the bowl center and vice-versa. This concept provides very efficient homogenization, even at low speeds. Its combination of stiff stirring members, tubes, arcuated blades and helices creates an extremely rigid framework allowing for thin precious metal wall thicknesses.

HOMOGENEITY IS QUALITY REQUIREMENT

One of the toughest challenges in glass making is achieving homogeneous and uniform properties of the glass melt. It is so important because it determines the quality of the final glass product and it is so difficult because there are so many properties that have to be created homogeneously. Tremendous efforts are spent on the subject of compositional glass homogeneity in the design and operating methods of glass melting furnaces. All of the process technology from batch melting over complete sand dissolution, avoidance of refractory contamination to bubble removal (fining) aims at creating a glass melt that has a uniform composition in every volume part.

Unfortunately, the homogeneity thus achieved deteriorates in the cooling and feeding channels after the furnace, simply because the glass melt acquires a non-uniform temperature distribution or material contamination. The undesirable results of such thermal and compositional inhomogeneity that are visible in the final glass product are known as cord, knots and striae (shown in Figure 1).

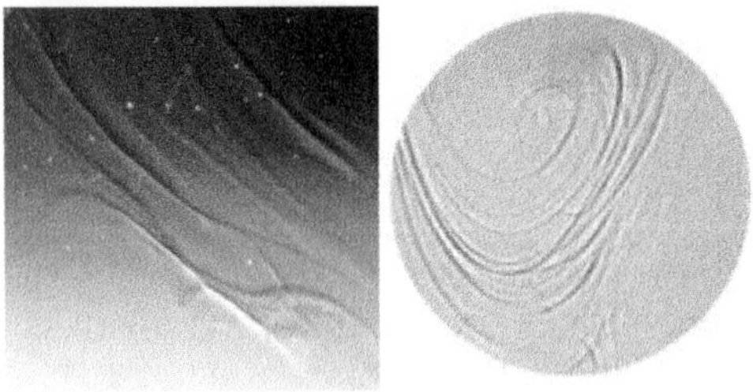

Figure 1. Knots and Striae Derived from Thermal and Compositional Inhomogeneity

POST REFINING HOMOGENIZATION BY STIRRING

Glass processes where cord in the product are unacceptable rely on stirring the glass melt before feeding. The stirring vessels are channels, crucibles, forehearth bowls, (coloring-)feeders and precious metal stirrer cells. A study of the existing stirrer executions reveals a large number of different stirrer shapes, a lot of them being proprietary to the inventors and users. There are tubular blades, blades with polygonal cross-sections, stirrers with central displacers, straight as well as curved blades (shown in Figure 2). Stirrer materials are refractory, molybdenum and platinum-rhodium alloys.

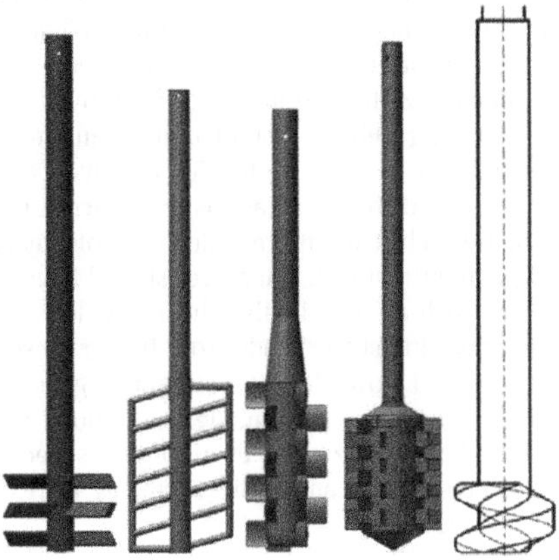

Figure 2. Examples for Different Glass Stirrer Shapes

This multitude of different stirrer designs triggers several questions, e.g., Which stirrer geometries provide efficient glass homogenization? Do all of the stirrer shapes follow fluid dynamic considerations or rather material constraints? What exactly makes a stirrer perform well? Analyzing the type of glass flow during stirring is one of the first fluid-dynamic considerations that helps to answer some of these questions. Any stirring of glass melts in every application takes place under laminar flow conditions. The absence of turbulence makes homogenization very difficult. The Reynolds number in a stirring setup indicates the type of flow, which is described by the following equation:

$$Re = \frac{\rho \cdot N \cdot D^2}{\eta} \quad (1)$$

Turbulence occurs at Reynolds-numbers greater than 10,000. An example for a typical stirring situation results in a Reynolds-number of Re=0.08, which means strictly laminar flow that is nowhere near turbulence. The glass viscosity at any reasonable stirring temperature is simply too high for that.

$$Re = \frac{2500kg \cdot 20 \cdot 0.14^2 m^2 \cdot m \cdot s}{m^3 \cdot 60s \cdot 200kg} \Rightarrow Re = 0.08 \quad (2)$$

This means that cord cannot be dispersed but only be stretched and folded by stirring (shown in Figure 3)!

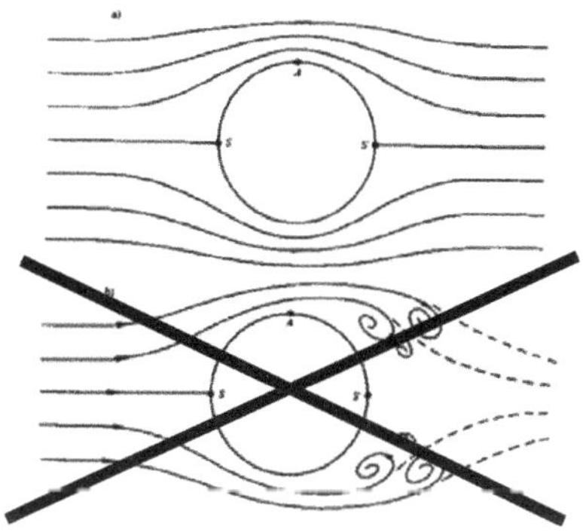

Figure 3. Stirring Takes Place at Laminar Flow

DEFINITION OF A DEVELOPMENT PROJECT

As most precious metal stirrers operate in cylindrical, vertical stirrer cells the task was set to study the fluid-dynamic implications of stirring inside such a cell and develop a homogenization device fulfilling the following criteria:

- Very high homogenization efficiency
- Self-supporting, Platinum Group Metals (PGM)-only, hollow design with lowest possible precious metal weight
- Very high operational safety
- Modularity to be adaptable for different stirring situations
- Longevity in operation

The product development was carried out successfully at Umicore Platinum Engineered Materials and led to the filing of a patent for the so called ARC*twister*. The development is described in the following paragraphs.

FLUID DYNAMIC ASSESSMENT BY MEANS OF PHYSICAL AND MATHEMATICAL ANALYSIS

The physical model emulates glass homogenization in a cylindrical, vertical stirrer cell in a continuous glass production. The model (shown in Figure 4) is constructed almost entirely out of acrylic plastic and pumps a clear fluid from the upper stilling basin through the horizontal inlet tube into the top of the vertical cylindrical stirrer cell. On its way down the stirrer cell the fluid must pass the stirrer and leave the vessel through the horizontal outlet tube at the bottom The outlet tube leads into the lower stilling basin from which the fluid is pumped up to repeat the cycle. All parameters of the model are set to have geometric and fluid dynamic similarity with the real application. That is to say that every parameter in the model-setup is scaled according to the rules of dimensional analysis to correlate with the real conditions. That is why the fluid was chosen to be a glucose solution as its viscosity can be varied with the sugar content. A variable drive above the stirrer cell allows stirrer models to rotate at different speeds and the glass level can simulate variable throughputs. The inflow tube allows the controlled injection of a colored tracer liquid at any position of the tube's cross-section. The colored tracer liquid is kept at the same viscosity as the rest of the fluid.

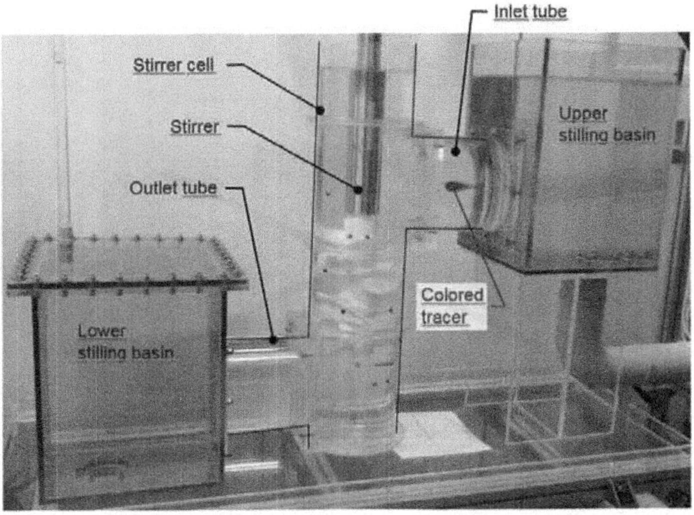

Figure 4. Physical Fluid Model Setup

As the colored tracer passes through the cell it is stretched, folded and deformed by the rotating stirrer. Different injection locations on the inlet tube cross section lead to different results. The homogenization behavior of nine injection points distributed across the inflow tube cross section was investigated. The center tracer trajectory (shown in Figure 5) proved to be the least affected by the stirrers, because the center pipe flow has the highest velocity resulting in the shortest residence time. It also hits the stirrer in the shaft area where its stirring efficiency is lowest.

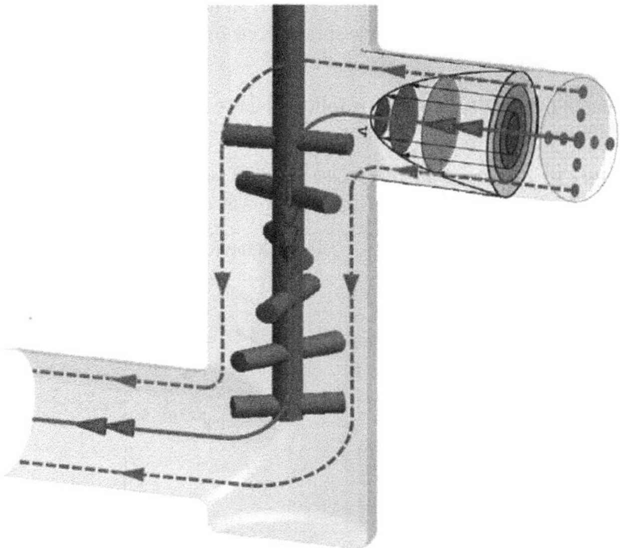

Figure 5. Center Trajectory - the most critical of the nine tracer points

The homogenization performance of most known stirrer geometries was tested and compared with this method. The stirrers' revolution speed was increased gradually until the tracer in each of the nine trajectories was stretched and folded to complete invisibility in the outlet tube. It was possible to make the tracer invisible with each of the tested shapes, however, the less efficient stirrers relied on high rotation speeds resulting in high torques. High torques require more precious metal in order to provide a stronger stirrer structure. Another important argument for slow rotation is the reduction of blade-induced shear stress in the glass melt. Low shear stresses in the glass melt help avoid cavitation which can be the source of mechanical reboil. Some processes also require low shear stress at the metal glass interface as metal erosion is detrimental to the product quality.

The performance is therefore defined according to the following rule: The stirrer that makes the center stream tracer disappear at the lowest revolution speed is the most efficient! The result was a stirrer classification from very efficient geometries to rather poor designs. The physical simulation is complemented by CFD-analysis. Instead of following individual tracer trajectories, the computational model uses a homogenization index to describe efficiency. The CFD-model was developed parallel to the physical glucose model and is very consistent with reality. With an established CFD-model the development, comparison and modification of stirrer shapes is very fast. It also allows the analysis of the mechanical stress distribution in the stirrer, which is important for the precious metal saving design of a stirrer.

THE ARCTWISTER CONCEPT

All high performance stirrers apply, at least partly, the design principles that are recommended for stirring at high viscous laminar flow. Those principles are:

- Agitation close to the cell wall to improve thermal homogeneity
- Increased glass residence time in the stirring zone by axial stirring effect.
- Frequent, repeated agitation by the blades.

The development of the ARC*twister* strictly followed these principles while incorporating several other innovations and features.

The basic idea is to combine three very efficient stirrer blade elements to an innovative concept that provides the highest homogenization performance. Those blade elements are:

Description	**Blade-element**	**ARCtwister**
ARC • Tilted arcuated blades that operate close to the cell wall. • Providing horizontal mass exchange • Acting as form fitting struts to distribute mechanical stress in the precious metal.		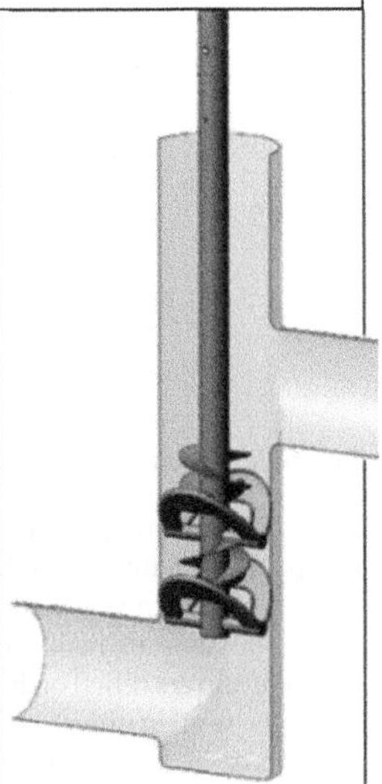
TUBULAR BLADE • A proven, very stiff design element • Providing repeated glass agitation • Carriers for the arcs	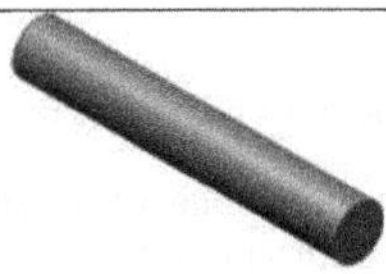	
HELIX • A proven, very stiff design element • Providing vertical flow • Increasing glass residence time		

The ARC*twister*'s functionality in terms of glass agitation is unique. The arcs push the glass towards the cell wall as well as pulling it towards the center of the cell. This horizontal mass exchange is combined with a vertical agitation by the helices and creates a three-dimensional flow. The vertical and horizontal mass transport ensures that the complete glass volume is constantly and repeatedly agitated by the tubular blades and the arcs. As the dynamic quality of this complex three-dimensional flow cannot be shown other than with moving images, the graphs in Figure 6 provide a simplified description of the glass transport.

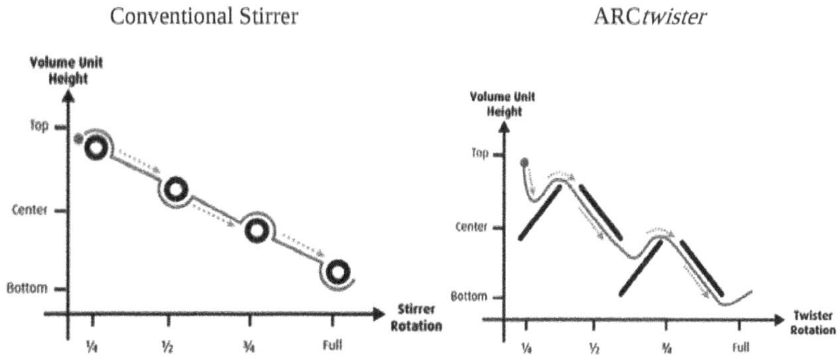

Figure 6. Conventional Glass Stirrer versus Intense Vertical Effect of ARCtwister

Conventional stirrers agitate the glass melt only in the area of the blade. At creeping laminar flow, the glass flows around the blade and receives only a mild homogenization. The volume around the blade is stretched, at best it is also folded. The glass stays almost completely unaffected in zones where the blade is temporarily absent. It continues to flow down the stirrer cell calmly until it is hit again by the next blade. At laminar conditions it hardly makes a difference whether the blades are shaped cylindrical, rectangular or like a paddle.

The ARC*twister* improves this behavior in several ways. The large tilted arcs constantly and gently guide the flow covering a big volume. They force the flow upwards and at the same time downwards depending on their slope. They also draw the glass from the cell wall towards the center, which is very beneficial for thermal homogeneity. At the center the glass is again transported vertically to increase residence time and increase the blade impacts per glass volume unit. A bottom view of the ARCtwister shows that the structure leaves no gaps for the glass to pass the stirrer unaffected (shown in Figure 7). Even if the ARCtwister was stationary it would still homogenize the glass similar to a static mixer.

Figure 7. Bottom View of the ARC*twister*

With these features, the ARC*twister* was tested and made the colored tracer disappear at a revolution speed of as low as 8 rpm. Other stirrers that are currently in actual glass systems need speeds of 18 rpm to 40 rpm to achieve the same result.

Just as important as the homogenization efficiency are the other design considerations that make the ARCtwister unique. The ARCtwister is a self-supporting design made out of oxide dispersion strengthened FKS®-alloy. Only the material's high creep resistance makes it possible to construct a device with a shaft and tube-blades that are hollow and seamless. The hollow tubular blades protrude through the shaft and interlock with the arcs. These form-fits relieve the stress on the welding seams. The arcs and the tube blades also form a statically rigid structure where the arcs support the tubes and vice-versa. In summary the ARCtwister features are:

- A low revolution speed for maximum homogenization
- Creep resistant FKS® -alloy
- Torque is conducted by interlocking elements making welds non-load bearing.
- Arcs and tubes form a static truss.

These characteristics make it not only fluid dynamically efficient but enable a lightweight design. All the precious metal wall thicknesses from the shaft to the blades can be reduced because the torque is low, the material is strong, the welds are only sealing and both the truss and the double helix are intrinsically rigid. The ARCtwister concept is also modular. The arcs, blades and helices can be adapted in number, size, orientation and characteristic to suit multiple applications, as shown in Figure 8. A lightweight ARC*twister* can quickly be customized to replace a less efficient stirrer in an existing vessel.

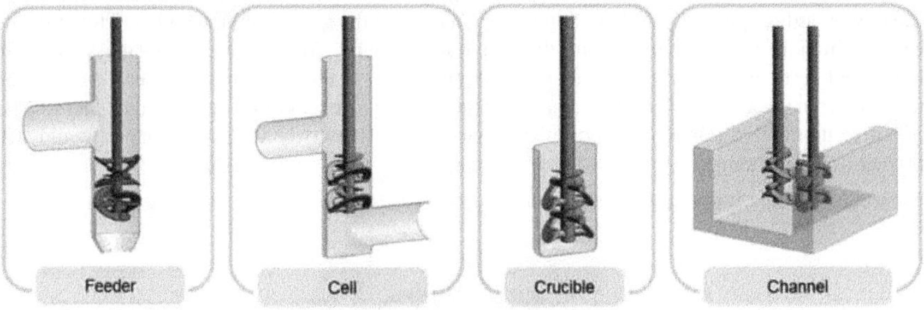

Figure 8. Various ARCtwister Designed for Specific Applications

Despite the ARCtwister's three-dimensional mass exchange characteristic it has no speed dependent impact on the glass level in its standard neutral configuration. However, it is possible to re-configure the three blade elements in such a way that the ARC*twister* has a conveying characteristic. Then, it can pump or suck the glass melt depending on the direction of rotation.

CONCLUSION

The ARC*twister* is an example for a successful product development. It is the intelligent combination of several technical disciplines that turns a good idea into a functioning product. Physical and mathematical fluid dynamic analysis leads the way to an efficient geometry. The fine grain stabilized FKS®PtRh-alloy allows the designers to create these shapes that would be otherwise impossible. Clever design ideas make the ARCtwister rigid, lightweight and modular. Finally, it is the advanced manufacturing techniques such as seamless tube drawing and precious metal welding that enhance operational safety.

NON-ISOTHERMAL GLASS MOULDING OF COMPLEX LED OPTICS

H. Kreilkamp, A. T. Vu, O. Dambon, ND F. Klocke
Fraunhofer Institute for Production Technology
52064 Aachen
Germany

ABSTRACT

The advantages of LED lighting, especially its energy efficiency and the long service life have led to a wide distribution of LED technology in the world. However, in order to make fully use of the great potential that LED lighting offers, complex optics are required to distribute the emitted light from the LED efficiently. Nowadays, many applications use polymer optics which can be manufactured at low costs. However, due to ever increasing luminous power which goes along with increasing temperatures and UV radiation, polymer optics reach their technological limits. Due to its outstanding properties, especially its temperature resistance, resistance against UV radiation and its long term stability, glass is the alternative material of choice for the use in LED optics. However, growing demands for complex yet low-cost glass optics pose a major challenge for glass manufacturers. Companies are continuously being confronted with the challenge arising from the trend towards increasingly complex geometries, ever higher levels of precision combined with a fast changing market and immense pressure on market prices.

This research is introducing a new replicative glass manufacturing approach, namely non-isothermal glass molding (NGM) in which glass preforms are reheated and molded into its final shape. New, flexible mold concepts are introduced to account for the optics' ever increasing complexity. Besides, alternative mold materials in combination with ultra-precision mold manufacturing technologies, namely ultra-sonic assisted diamond turning are used to guarantee highest form accuracy of the used molding tools. The key to success for a fast time to market is the integration of FEM simulation at the early stage of the process development, which will also be presented in this research. This thermos-mechanical process modeling provides a systematic approach for glass preform design, mold compensation, and process optimization. Finally, a business case including machine concepts is presented for competitive serial production from 10.000 up to 30 Mio. optics per year. In conclusion, short process times in conjunction with the elimination of any requirement for finishing work to be conducted on parts, have made NGM the technology of choice for serial production of complex LED glass optics.

INTRODUCTION

According to a study undertaken by the U.S. Department of Energy, the unit sales of LED lamps will more than double in the next four years [1]. By 2020 a total of 1.6 billion LED lamps will be sold per year. Due to their energy efficiency, long service life and enormous flexibility in terms of lighting form factor an enormous transition in the lighting market is observed. In tandem with the number of LEDs, the requirement for LED optics is projected to increase. However, as the sales of LED lamps increase, the market pressure in terms of manufacturing costs is growing as well. As Figure 1 illustrates, the relative manufacturing costs for LED optics are projected to be reduced by 60% from 2014 – 2020. This is putting an enormous pressure on optics manufactures for cost-efficient manufacturing.

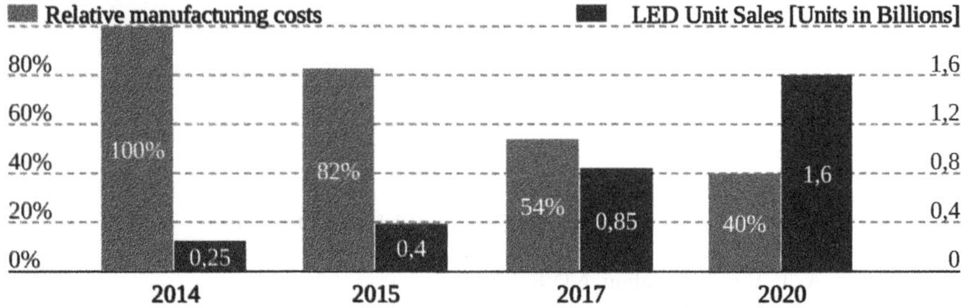

Figure 1. Cost Projection for LED Optic (based on [1])

Currently, the most cost-efficient way of manufacturing LED optics is by plastic injection molding. For a wide range of applications, polymer optics offer many advantages. They are very light-weight and can be molded into complex shapes, including multiple optical surfaces and filigree mounting features. However, the use of polymer also induces several limitations. High UV radiation leads to significant degradation of the polymer, also known as the yellowing or browning effect [2]. Furthermore, polymer optics show only a very limited thermal resistance.

In order to be more cost efficient, a trend towards LEDs with even brighter light heads, which offer the ability to substitute multiple less bright LEDs, can be observed. As a consequence of this trend optics have to withstand higher thermal loads. Since polymer optics have a very low thermal resistance compared to glass, polymer optics can only be used to a limited extend. In comparison to polymers, glass offers some unique properties like an excellent heat and UV resistance, great chemical resistance and high scratch resistance which are vital for future LED applications.

Whereas there are already a number of established and economically efficient manufacturing processes suitable for plastic LED optics, the manufacture of glass optics for high-end applications is still associated with high levels of outlay. To overcome this lack, a new non-isothermal glass molding technology has been developed to manufacture complex, high precision LED optics at polymer competitive prices.

CONCEPT OF NON-ISOTHERMAL GLASS MOULDING

Developed in the 1990th, precision glass molding has become an established technology for series production of precise glass optics [3]. In precision glass molding a well-defined glass preform is loaded into a precision molding tool at room temperature. The glass and the mold are then heated up together by infrared heaters. Once a specific molding temperature is reached, the glass preform is deformed into its optical shape by applying a defined force onto the molds. As soon as the glass is completely deformed, the glass and the mold are cooled down in parallel. By applying a defined holding force during the cooling step, thermal shrinkage of the glass can be compensated to some extend resulting in highly precise optics. Once, the glass is cold enough, the molds open and the optic can be unloaded.

Since the entire molding tool needs to be heated up and cooled down for each process, the cycle time is usually around 15 – 20 minutes (see Figure 2). During the whole time, the glass is in contact with the mold leading to high mold wear and therefore limited mold lifetime.

Combined with relatively expensive molding tools, the precision glass molding process is therefore not suitable to reach the cost requirements of LED optics.

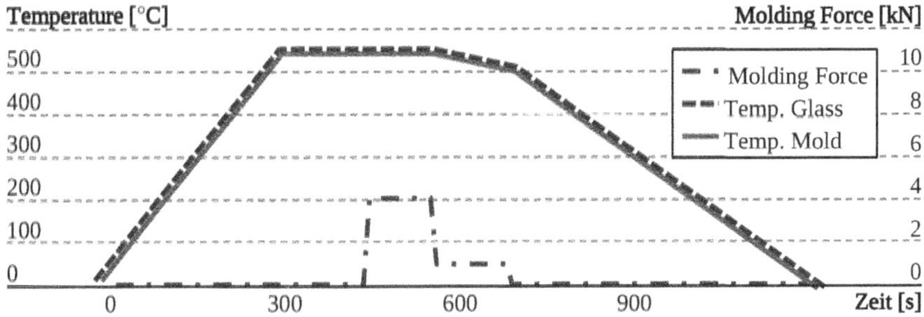

Figure 2. Concept of Precision Glass Molding (Isothermal)

Therefore, a non-isothermal glass molding process has been developed at Fraunhofer IPT to meet the dual challenge of high precision and complexity at low manufacturing costs. In non-isothermal glass molding a glass preform is loaded onto the lower mold at room temperature. The glass is then heated up in an external heating oven (see Figure 3). Since the heating oven is constantly at high temperature, the heating of the glass preform is much faster. Furthermore, several glass preforms can be heated up in parallel.

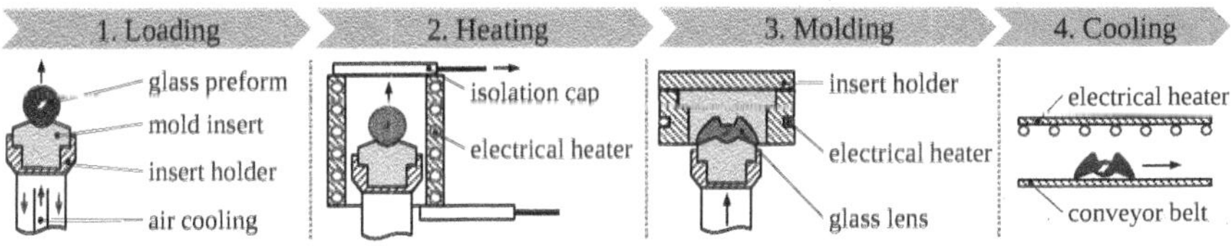

Figure 3. Process of Non-Isothermal Glass Molding

Once the glass is at sufficient molding temperature, a servo-electric axis moves the lower mold upwards through the heating oven towards the upper mold, which is held at constant temperature significantly lower than the molding temperature (see Figure 4). The glass is molded into its final optical shape. Compared to precision glass molding, the molding time is significantly shorter since the heat of the glass is transferred into the comparatively colder mold very quickly. Typically, the molding time is around 10-15 seconds. As a consequence, the molds do not suffer from long heating and cooling cycles. Since the molds' temperature can be kept constantly low over many processing cycles, it results in substantially prolonged mold service life.

After the glass is being molded, the glass optics are gradually cooled down to room temperature in an external annealing lehr. During this cooling process, stresses in the glass introduced by the fast cooling during the non-isothermal molding, are released during this process. Typically, cooling times are between 60 and 180 minutes depending on the glass type and geometry.

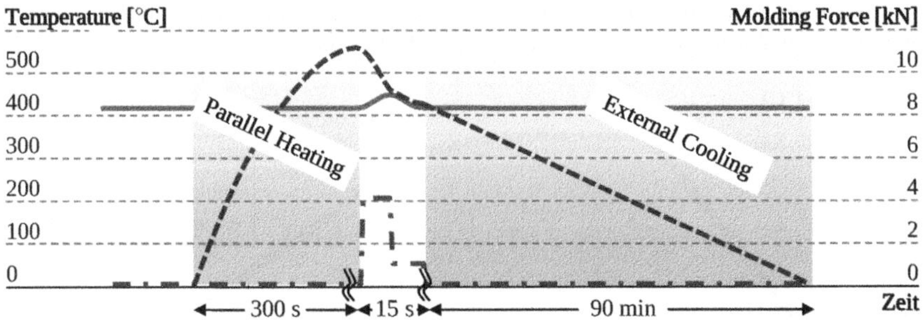

Figure 4. Concept of Non-Isothermal Glass Molding

In summary, the short process times for molding along with the fact that no further steps are required for finishing makes non-isothermal glass molding ideal for large-scale manufacture. Due to the short contact time between glass and mold, steel can be used as mold material. This allows the use of efficient mold manufacturing technologies, especially diamond turning which is a suitable technology for precise mold manufacturing [4]. Diamond turning offers a wide geometrical flexibility which makes it ideal for the manufacturing of complex freeform molds for LED optics replication [5].

The advantages of glass as an engineering material can be exploited especially in LED optics. However, the LED market is a very dynamic, fast changing market. The time-to-market for new optical designs is typically only several months long, so that the development of non-isothermal glass molding processes needs to be carried out with highest efficiency. A deep understanding of the underlying mechanism of glass molding is the basis to develop sufficient glass molding processes for new LED optics in short time. FE-Simulation is the key to success for a fast development and ramp-up process.

INTEGRATION OF FEM-SIMULATION INTO THE PROCESS CHAIN

Although intensive process know-how about non-isothermal glass molding has been gained over the years, the process development for new optical designs is still very time consuming. An optimal preform must be designed for each new glass optic, the molding tools need to be designed, and the optimal process parameters must be identified. During non-isothermal glass molding the glass undergoes large temperature changes, leading to high shrinkage of the glass. Typically after the glass has been molded, the form accuracy will be measured. Based on the form deviation caused by the high shrinkage of the glass, the mold geometry needs to be recalculated and remanufactured accordingly. Finally, the remanufactured mold will be tested again and the remaining form accuracy will be measured again (see Figure 5). This iterative approach is performed until the remaining form accuracy is within the desired specifications. Based on the lens specifications, up to three iteration cycles might be necessary. Therefore, this approach is very time- and cost intensive and contradicts the requirement of a fast changing and dynamic LED market.

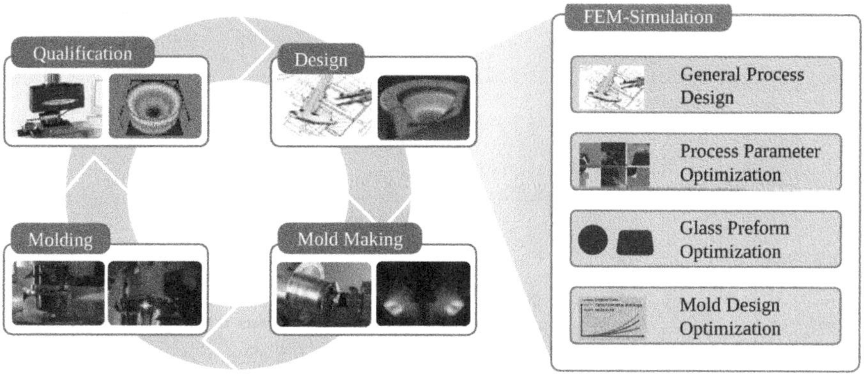

Figure 5. Integration of FEM Simulation into the Process Chain

Therefore, an FEM-based Simulation tool has been developed to reduce the development time and associated costs. Numerical modeling in glass molding however requires a complex thermo-rheological-mechanical model in order to predict the complex behavior of glass during the entire manufacturing process. Especially the temperature distribution within the glass strongly influences the final deformation of molded optics, the optical properties and the residual stress. Thus, an accurate thermal modeling is highly crucial for the prediction of glass temperature. This research performs a numerical thermal model, taking into account three heat transfer mechanisms, i.e. conduction between glass and mold, convection among glass, molds and environment, and radiation are taken into account, as described in Figure 6.

For mechanical modeling, the viscoelastic behavior of the glass during the molding and annealing stages are the most critical influences to predict the final shrinkage of the molded glass. In non-isothermal glass molding, the glass preform is heated up to temperatures above the transition temperature (Tg) at which the viscosity of the glass lies in the range between 107.6 to 109 dPas. At this temperature, the glass strongly behaves as a viscoelastic material, which can be described by a generalized Maxwell model [6]. The viscoelastic behavior of glass is also highly temperature dependent. This temperature-dependent behavior is characterized by Thermo-Rheological Simple (TRS) behavior and is modeled by using WLF shift function [7].

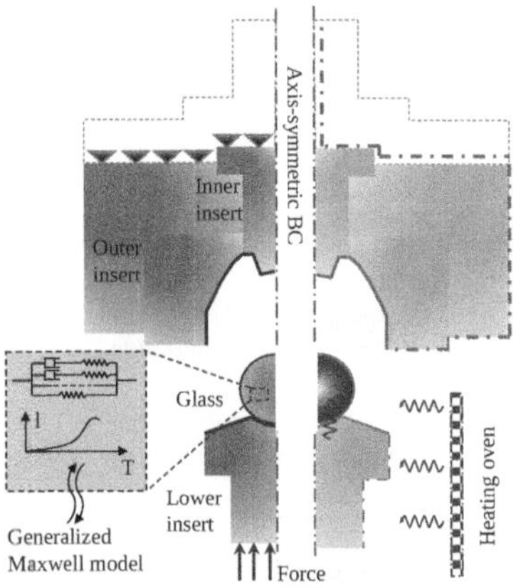

Properties	Suprax 8488 glass	1.2787 steel mold
Thermal conductivity, k_c (W/m°C)	1.2	25
Density, ρ (kg/m³)	2300	7700
Specific heat, C_p (J/kg°C)	750	480
Elastic modulus, E (MPa)	67×10^3	210×10^3
Poisson ratio, ν (-)	0.2	-
Thermal expansion coefficient, α (/°C)	4.1×10^{-6}	10.1×10^{-6}
Transition temperature, T_g (°C)	554	-

values at room temperature

Visco-elastic behavior · Elastic behavior · Volumetric conduction
Structural contact · Thermal contact · Convective surface · Heat radiation

Figure 6. Boundary Conditions of FEM-Simulation (left) and Material Properties (right)

RESULTS

Figure 7 shows the mold filling behavior as it has been simulated with above mentioned method using the software ABAQUS. Starting from a spherical preform the glass deforms with 5 seconds into its final shape. However, the glass is not filling the lower corner of the mold completely since the glass temperature near the surface is not sufficiently high in this area. The temperature in this area is only around 650°C which corresponds to a viscosity of η = 1010,1 dPas. Under the given load of F = 8 kN the glass does not further fill the cavity leading to a residual deviation of the optic.

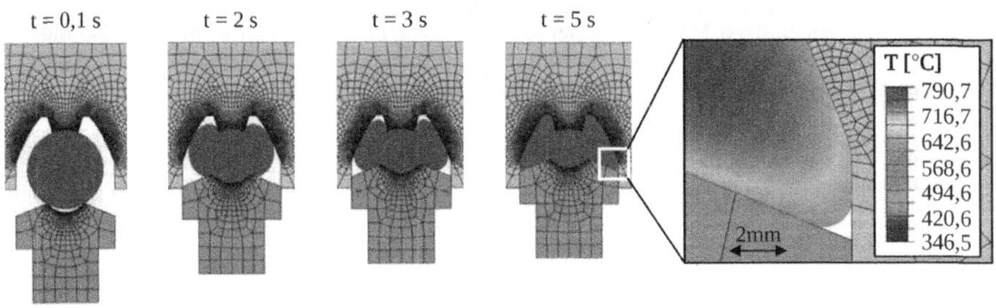

Figure 7. Simulation Results of Mold Filling Behavior

In a second step, molding experiments have been conducted to validate the simulation results. The experiments have been performed on a linear press "Füller GT52" with servo-axis. Afterwards the glass optics have been measured to analyze the overall form deviation using an optical 3D scanner "Steinbichler Comet 5". As Figure 8 illustrates, the measured geometry is in good agreement with the geometry obtained by means of FEM-Simulation.

Process Parameters

Mold Material: 1.2787
Glass: SUPRAX 8488

T_{Mold}	=	500°C
$t_{heating}$	=	168 s
T_{oven}	=	1000 °C
v_{max}	=	20 mm/s
F_{max}	=	8 kN

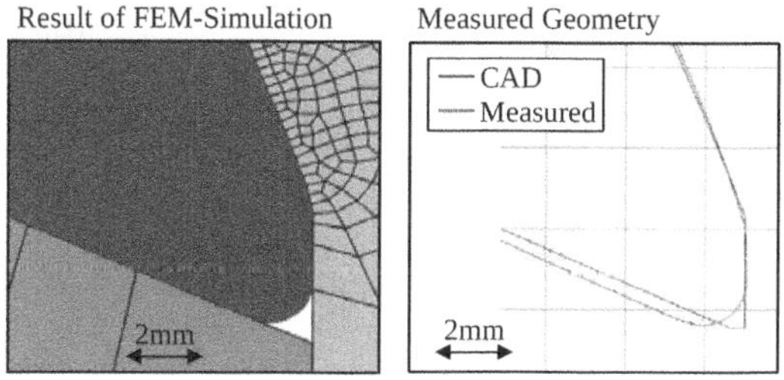

Figure 8. Comparison of FEM Simulation Results with Experimental Data

Based on the simulation results an optimization of process parameters has been performed in order to optimize the mold filling behavior. Several simulations have been run with varying process parameters and the effect on the filling of the lower corner has been analyzed. Finally, an increase of heating time from theating = 168 s to theating = 174 s was leading to complete filling of the mold. The glass optic molded with the optimized parameters obtained from FEM-Simulation is shown in Figure 9 in comparison to the initial result. Again, the experimental result in terms of mold filling behavior is in good accordance with the simulation data.

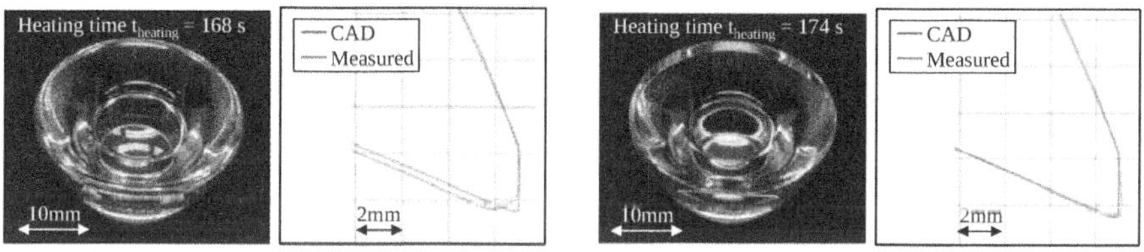

Figure 9: Optimization of Mold Filling

For further in depth analysis of the form accuracy of the optical surfaces a profilometer "Taylor Hobson Form Talysurf PGI 1250" has been used. Furthermore, the surface roughness has been analyzed using a white-light interferometer "Bruker Contour GT-K". The overall form deviation of the measured optical surface is less than P-V = 4 µm with surface roughness of Ra < 5 nm (see Figure 10). Furthermore, the repeating accuracy has been tested by molding 10 lenses with identical process parameters. The form accuracy of all 10 lenses has been measured revealing that the standard deviation of the resulting form accuracy is less than 0,46 µm. These results demonstrate that non-isothermal glass molding is not only capable for the mass production of lighting optics but can also be used for applications requiring for higher form accuracies, e.g. for the manufacturing of sensor optics.

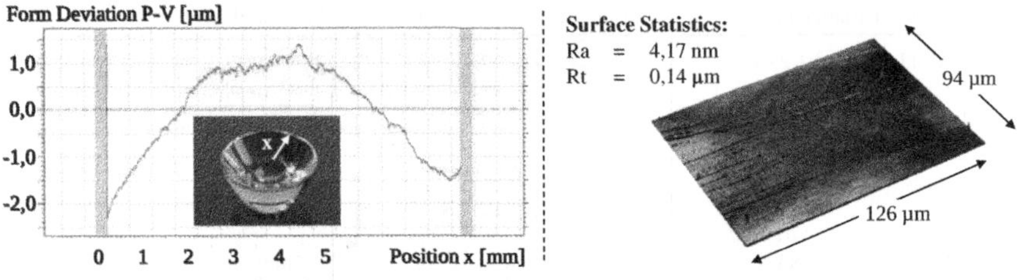

Figure 10. Reachable Form Accuracy (Left) and Surface Roughness (Right)

SUMMARY AND OUTLOOK

A non-isothermal glass molding technology for the cost-efficient manufacturing of complex lighting optics has been developed successfully. The non-isothermal glass molding is capable to solve the contradicting requirements of high precision coupled with low manufacturing costs. An FEM-based simulation model has been developed and successfully validated in molding experiments. The mold filling has been optimized with the help of FEM-Simulation and the resulting form accuracy was below P-V = 4 µm with surface roughness below Ra = 5 nm.

For large scale industrial application of this process a business case has been developed together with a big lighting manufacturer and the machine supplier Füller Glastechnologie. The business plan reveals that production of 10.000 up to 30 Mio. optics per year is possible. In comparison to the injection molding of plastic optics, the calculations indicate that competitive price of the glass optics can be reached if a high machine utilization of min. 80% can be realized. In conclusion, the non-isothermal glass molding technology has a wide range of potential applications across many photonic sectors especially in large volume production of complex but low-cost glass optics.

ACKNOWLEDGEMENTS

The authors gratefully acknowledge the financial support of the European Commission for the support of this research within the project “CENTiMO” (FP7-SME-606105).

REFERENCES

[1] N. Bardsley, S. Bland, D. Chwastyk, Manufacturing Roadmap - Solid-State Lighting Research and Development, U.S. Department of Energy, 2014.

[2] L. W. McKeen, The effect of long term thermal exposure on plastics and elastomers, William Andrew, Waltham, 2014.

[3] A. Y. Yi et al., Development of a compression molding process for three- dimensional tailored free-form glass optics, Appl. Opt. 45 (2006) 6511–6518.

[4] E. Brinksmeier, R. Glaebe, J. Osmer, Surface Integrity Demands of High Precision Optical Molds and Realization by a New Process Chain, Procedia Engineering 19 (2011) 40-43.

[5] C. Brecher et al., Technical advances in ultra-precision machining of freeform optics by function-oriented testing, Proc. of the Sixth Int. Conf. on Manufacturing Research 12 (2008) 73–80.

[6] F. Wang et al., Numerical simulation assisted curve compensation in compression molding of high precision aspherical glass lenses, J. Manuf. Sci. Eng. 131 (2009) 011014.

[7] M. L. Williams, R. F. Landel, and J. D. Ferry, The temperature dependence of relaxation mechanisms in amorphous polymers and other glass-forming liquids, J. Am. Chem. Soc. 77 (1955) 3701–3707.

Refractories

TIN BATH BLOCK EVOLUTION AND DEVELOPMENT: A CASE OF TOIL ENDEAVOUR AND FEARS

Christopher J Windle
Technical Manager
DSF Refractories & Minerals Ltd
Derbyshire, United Kingdom

ABSTRACT

Since the advent of the float glass process devised by Sir Alastair Pilkington in the early 1950s, the performance of the tin bath has been crucial; indeed fundamental to product quality. Profitable exploitation of the float glass process commenced in the early 1960s, however ever increasing campaign objectives and production output over the following decades imposed high demands on the tin bath blocks. The problems encountered with the tin bath block were all unique; they required extensive investigation, novel thinking and a truly collaborative approach between the glass maker and refractory manufacturer. In addition, the kinetics of many undesirable effects were extremely slow taking years to become apparent, therefore the "fear" was that many baths could be affected by the same problem as block compositions were common to all. An inappropriate composition for the bath block can lead to a multitude of problems which result in lost production, and high cost issues. The right formula of bottom block will provide continuous trouble-free production for many years. This paper and presentation reviews the problems in a historic and current context and outlines the compositions available today; a result of continuous development since the 1960s. It also highlights where lessons have not necessarily been learnt; where rules have not been followed and ultimately determines that the "toil, endeavour and fears" of all the past material and process engineers should not be taken for granted.

INTRODUCTION

The float glass process revolutionized the manufacture of flat glass, it provided large, fired finished, high quality plates of glass for building and automotive applications far exceeding the capability of sheet and plate production not only in volume but also unit cost. Float glass manufacture has been exponentially successful accounting for 90% of the world's flat glass demand from over 400 float glass lines producing in excess of 50 million tonnes per year. From the early 'equilibrium' glass thickness of 6.8mm; thicknesses now range from ~0.4 to 25mm.

At the heart of the process is the "bath" developed over ~7 years in the 1950s. With overall dimensions ranging up to 80m in length by 8 m wide; this marvel of engineering can take viscous glass at ~1100°C allow it to spread under the influence of gravity and surface tension atop liquid tin a medium that allows a continuous ribbon to be pulled at a rate of up to 1000 tonnes per day. The glass exits the bath at around ~600°C; cold enough to resist marking as it is transferred to the lehr for temperature equalization/stress removal.

The molten tin is in contact with, and insulated by refractory blocks with the main sub-divisions termed as bottom and sidewall blocks. These blocks in turn are encapsulated in an electrically heated, air tight steel casing with a N_2/H_2 (N_2:- 90-95%, H_2:- 5 – 10%) atmosphere preventing tin oxidation (<5ppm O_2) . Figures 1 to 3 below show the interior of an operating bath and transverse and longitudinal schematic illustrations.

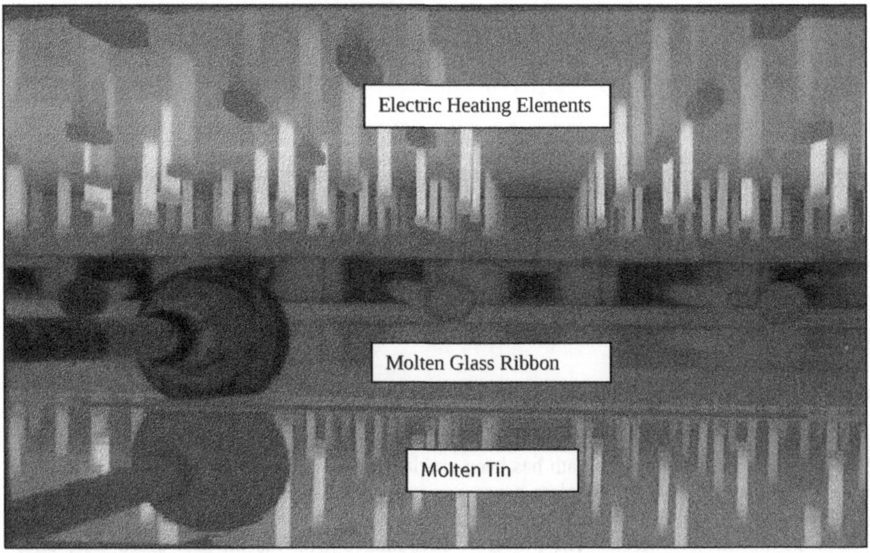

Figure.1. Float Bath Internal (Courtesy of GlasRef Consulting)

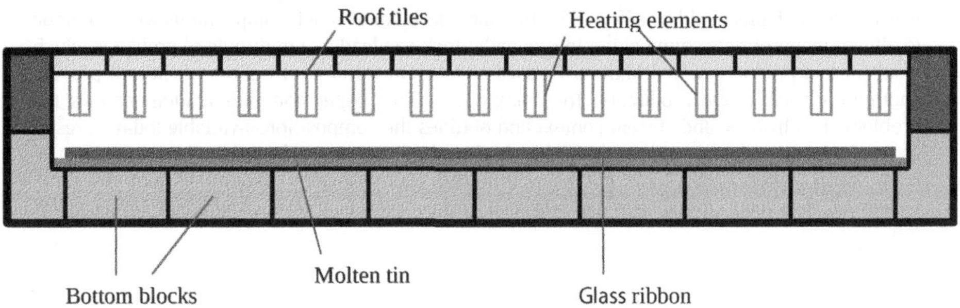

Figure 2. Float Bath Schematic (transverse section view)

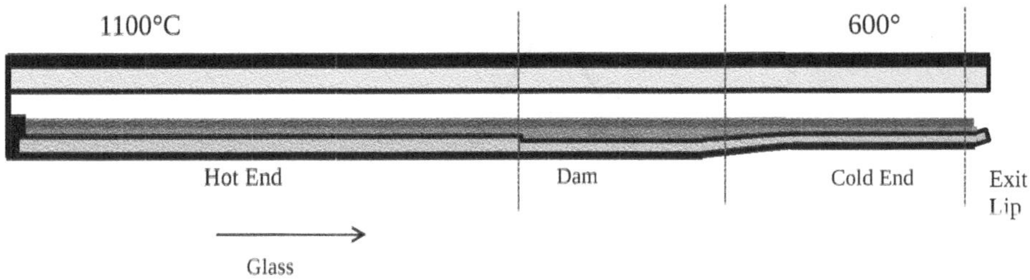

Figure 3. Float Bath Schematic (longitudinal section view)

HISTORY OF BLOCK DEVELOPMENT

EARLY DAYS

Originally (upto 1970), 25-30%Al_2O_3 blocks were used for bottom blocks, during the development phase of float glass there was no reference to the refractories in the bath as "giving trouble because of their composition" [1]. As with all new processes however there were truly unknown phenomena, the first of these was encountered when vitreous faults termed "tadpoles" sometimes stretching several hundred metres appeared attached to the underside of the glass ribbon. It was quickly determined that the composition of these faults was a combination of the block Al_2O_3-SiO_2 content and Na_2O, the ternary eutectic (lowest melting point) of a composition starting at 25-30% Al2O3 is 732°C, close to the albite phase field. The Na_2O component was the result of Na+ ions migrating from the ribbon through the tin scavenging O_2 from the block and consequently reacting with /absorbing into the block tin contact surface and cumulatively moving to liquid phase shown in Figure 4. The liquid phase created although termed viscous was a lower density than the molten tin and therefore floated upwards to become attached to the underside of the ribbon. As the ribbon movement is continuous the defect could be stretched to give the appearance of the head and tail of a tadpole; a very distinctive fault which was often the major cause for glass rejection. The electrochemical process resulting in the transfer of Na+ ions from the ribbon through the tin was extensively researched by Muller et al [2]. The solution to the problem was to use higher alumina blocks (~40%); thus avoiding the compositions associated with "tadpole" defects.

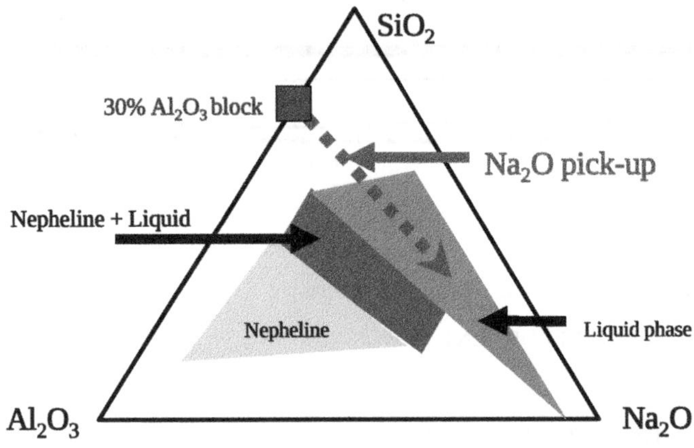

Figure 4. Simplified Isothermal Phase Diagram at 1000°C (Courtesy of GlasRef Consulting)

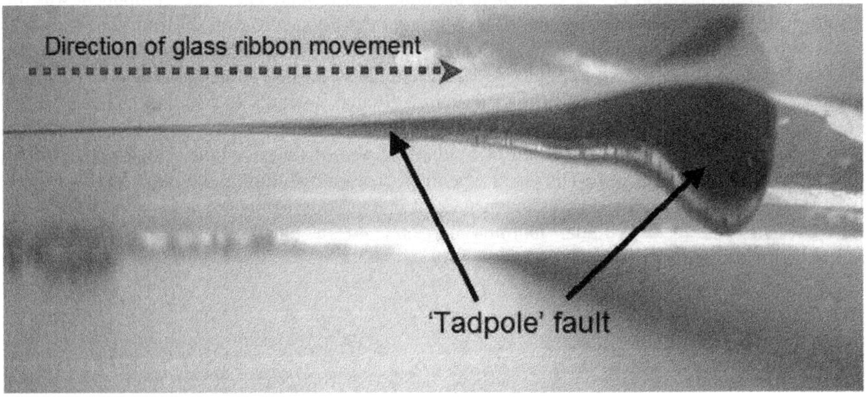

Figure 5. "Tadpole" Fault on the Underside of the Glass Ribbon (Courtesy of GlasRef Consulting)

THE UNEXPECTED BUBBLE

With tadpole defects now resolved no further problems were perceived until a major glassmaker in North America decided to use an indigenous bottom block supply. When this Pilkington licensee started up "their bath boiled" [1], a completely unexpected effect from a bath constructed from near identical materials to the established plants. The American bath block refractory met all the then known criteria, however as soon as glass contacted the tin surface there was a massive outbreak of bubbles on the underside of the glass ribbon (open bubble), resulting in a total loss of product. An extensive investigation commenced which involved considerable toil and bravery to fold the ribbon and thus encapsulate the problematic gas at the heart of the bubble. It was thus determined that the gas was H_2 enriched bath atmosphere. The initial challenge was to find the phenomenon at work; "transpiration" flow of gas fitted the observations; specifically different to diffusion in that the flow of gas is against the temperature gradient imposed i.e toward the higher temperature.

The driving force behind the gaseous flow known as "transpiration pressure" (Tp) is related to the temperature gradient through the block by the following equation:

$$Tp \propto (Tmax/Tmin)^{0.5} \quad (1)$$

However, the major component of the bubble defect was H_2 and therefore this molecule was somehow undergoing preferential transpiration. The answer lay in the pore size distribution of the refractory material with particular focus on the mean free path of hydrogen and nitrogen. Analysis of bath block samples in respect to pore size distribution showed that problematic material had ~75% of pores in the range of 0.1 to 0.6 μm (mean free path of a H_2 molecule is 0.17μm).

Figure 6 is a schematic illustration of how H_2 molecules move through a fine pore structure analogous to a ping pong ball moving through a tube; conversely large pores in the refractory structure "short circuit" the H_2 movement by creating a multi-faceted void within which the molecule "bounces" in multiple directions.

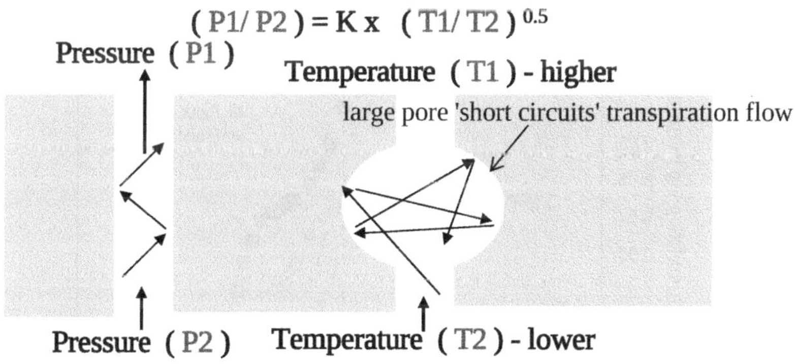

Figure 6. Schematic Illustration of Movement of H_2 Molecules (Courtesy of GlasRef Consulting)

The pore size vs transpiration risk is shown in Figure 7, fundamentally a material which has a high percentage of large pores is a low risk in regard to H_2 (or N_2) transpiration and therefore "open bubble defect". Figure 8 shows two materials (A & B) currently used in tin bath bottom block constructions. Although the chemistry and physical properties of these materials is similar the pore size distributions are distinctly different. Specifically material B shows a lower Hg intrusion and therefore a lower apparent porosity than A; the % of large pores >8μm in B is lower than A (permeability B < A), however B has only a few pores in the critical thermal transpiration range of 0.1 – 0.6μm.

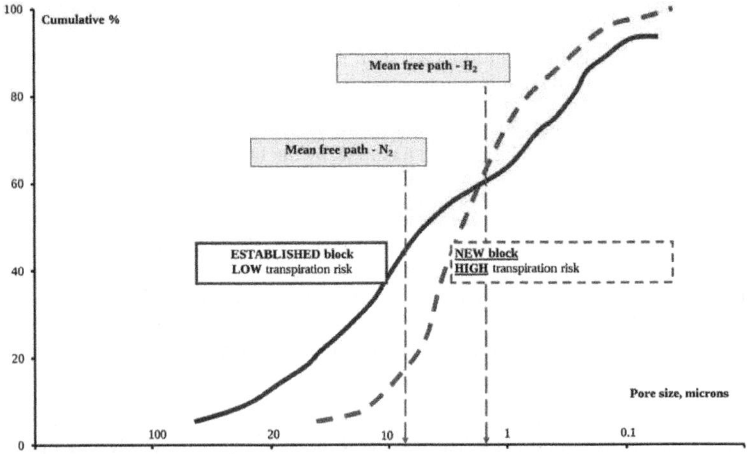

Figure 7. Pore Size vs. Transpiration Risk (Courtesy of GlasRef Consulting)

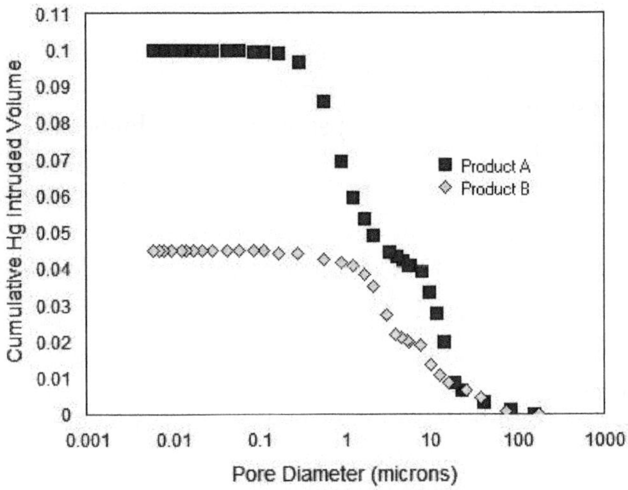

Figure 8. Comparison of Pore Size Distribution of Tin Bath Block Materials

Transpiration or in context thermal transpiration is a high temperature process; an ambient test was required so that significant and statistically relevant numbers of a batch supply could be assessed. Pilkingtons developed the diffusivity test and its relationship with transpiration pressure is shown in Figure 9.

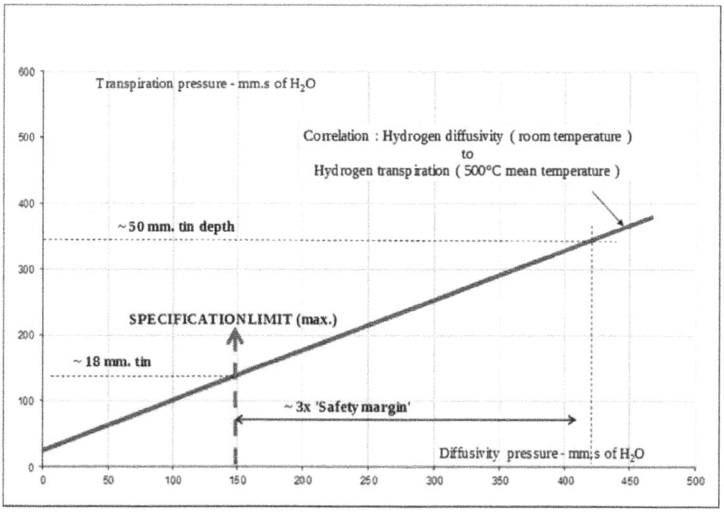

Figure 9. Diffusivity Pressure vs. Transpiration Pressurte Courtesy of GlasRef Consulting

Nearly all specifications for H_2 diffusivity state a maximum of 150 mm of H_2O; this provides a 3-fold safety margin in regard to transpiration/open bubble defects. In the American case a low firing temperature had resulted in a finer pore structure; the coarsening of porosity during sintering processes is well known and documented therefore with the "new" information firing temperature and duration could/can be adjusted to give an optimum pore size distribution. Although "open bubble defect" potentially should be confined to the annals; refractory manufacturers must not be complacent and must continue to test each batch of supply. In addition to the bath block, the securing rammable or castable presents a significant portion of the block volume (~3.5 %) and block surface (~3 %). Therefore, these should also be tested to ensure that they are not conduits to trapped H_2; tests on rammed material typically give ~45 mmH_2O, well within specification limits, whereas castables and specifically free flowing materials are designed to have the minimum amount of free space and consequently have a very fine pore structure, ideal for H_2 transpiration with values around ~600 mm H_2O; exceeding the threshold limit. To the author's knowledge this has not been a problem or least not identified as the source of a problem; nevertheless all components of the bath construction should be checked to be "H_2 safe".

STRAINING TO EASE THE STRESS

In the late 1970's a further problem arose when a tin bath experienced bottom block shear and detachment. This imparted concerns that other baths may be affected as the bottom block material was common to many operational lines. Indeed several float baths experienced a problem now known as "Seven-Inch Splitting". The bottom blocks were shearing in a horizontal plane with the detached part floating upwards through the denser tin. This disrupted the steady movement of the glass ribbon through the bath; the only option was to cool down and repair with an associated heavy commercial consequence. The problem root cause was the desire to keep the bath construction "tin tight" and therefore reversible expansion gaps between individual blocks were designed to be less than theoretical. This created a high compression zone which in turn produced a perpendicular tension force on the blocks. From analyses it was found that this force was highest at depths of 6 to 8 inches matching the typical shear depth (Figure 10).

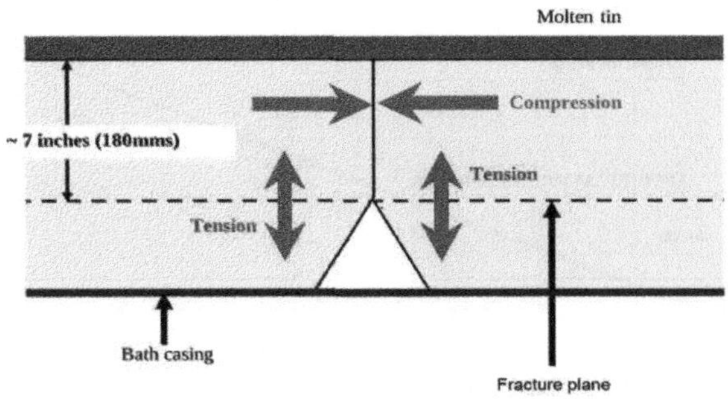

Figure 10. Schematic of Compression and Tension Forces on a Tin Bath Block (Courtesy GlasRef Consulting)

The failure mode was a translation of restricted expansion into a tensile force and therefore a material which could strain elastically to absorb stress (low elastic modulus, E) would be advantageous. This does not exclude a higher tensile strength (modulus of rupture M.O.R), hence an M.O.R/E quotient was devised. In conjunction with the quotient Pilkington's also developed a proof test (Figure 11) which was a small scale "mimic" of the compression and subsequent tensile force experienced by a full size block. Both the compressive load at failure and the deformation (strain) of the material at failure could be determined. The test was an excellent simulant of the observed phenomenon as failure always occurred at a position corresponding to the "Seven-Inch Splitting" effect.

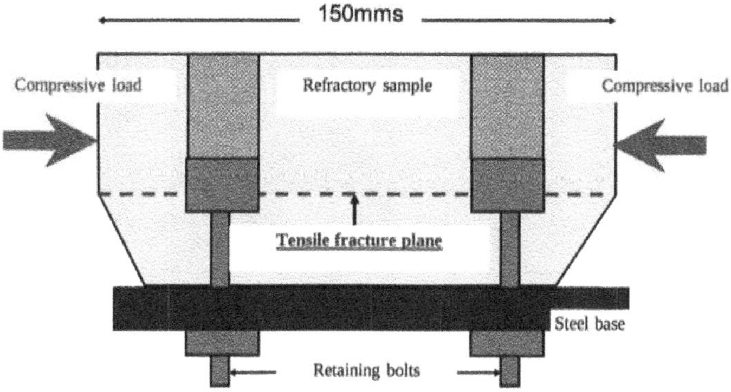

Figure 11. Schematic of Pilkingtons' Proof Test (Courtesy GlassRef Consulting)

Following testing of many different bottom block materials (A to E in Fiure 12); the supposition that materials with a higher ratio of (M.O.R/E) could withstand a higher compressive strain before fracture was confirmed: "Failure strain" α M.O.R/E. The 7" shearing of bottom blocks in several baths which were constructed from otherwise traditionally successful materials was the tracked down to small manufacturing modifications. These modifications had successfully improved the appearance and surface hardness of the blocks but inadvertently made them less ductile (A1 to A2, Figure 12). The problem was unexpected, however from a combination of excellent field observations translating into laboratory scale testing, it was possible to provide the refractory manufacturer with the criteria that should be met for thermo-mechanical behaviour of the material in the tin bath. Another "golden rule" had been established.

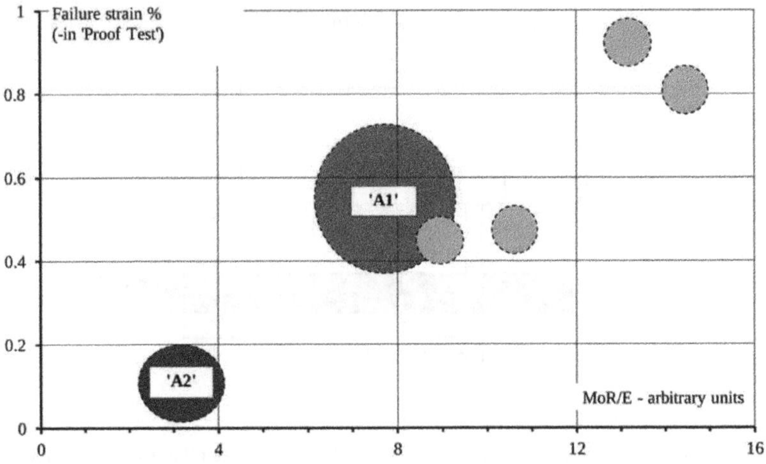

Figure 12. Courtesy of GlasRef Consulting

THE LAST HURDLE

Thermodynamics dictate that an alumina-silicate bath block material will form a feldspathoid with alkali at the appropriate temperature. In float glass the migration of sodium ions from the ribbon through the tin, scavenging of oxygen, and subsequent reaction results in the formation of nepheline. Nepheline formation results in volume expansion which creates a stress between the surface and bulk of the bath block subsequently leading to a "flake" (shown in Figure 13), which shears from the block surface, rises through the tin and damages the glass (show in Figure 14).

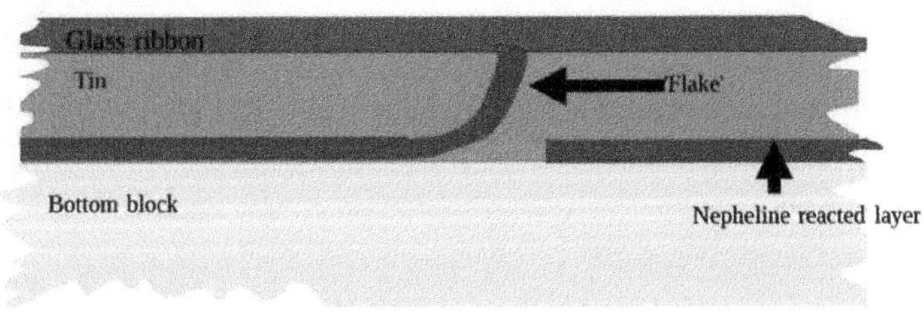

Figure 13. Schematic of Nepheline Flake Formation (Courtesy GlasRef Consulting)

Figure. 14 Nepheline Flakes Retrieved from a Tin Bath

The process has become known as “nepheline flaking” and although first detected in the 1980’s it is a current problem, and therefore illustrates that potentially the lessons of the past have not been heeded. When the problem first occurred it appeared that some materials were more susceptible than others; once again extensive testing determined that the flaking phenomenon was related to permeability and that attempts to reduce the elastic modulus and hence the 7” splitting had resulted in a more permeable material. The hypothesis that a permeable material would be susceptible to flaking seemed logical, as higher permeability would allow easier alkali penetration, resulting in a greater volume of affected refractory with heterogeneous behaviour to the bulk of the block. The change to established materials had been subtle (Table.1) however the change in permeability had been significant. To reduce the elastic modulus and therefore avoid 7” splitting the glass phase of “1” had been reduced which increased the permeability “2” and also reduced the capacity of the material to absorb Na_2O in an amorphous state rather than the expansive crystalline one of nepheline.

Table 1. Details of flakes formed (Courtesy GlasRef Consulting)

Refractory	Problem	$Na_2O + K_2O$	% Glassy Phase	Air Permeability nPm
1	Seven inch splitting	1.1 – 1.3	32 - 38	<2
2	Nepheline flaking	0.6 – 0.9	15 - 25	9 - 12

A test was devised to give an indication of the block's resistance to soda penetration; a small hole is drilled into a block sample filled with soda ash. This sample is then cycled between 800 and 1000°C every 2 hours for 7 days. On cooling the sample is sectioned to observe the degree of penetration as shown in Figure 15.

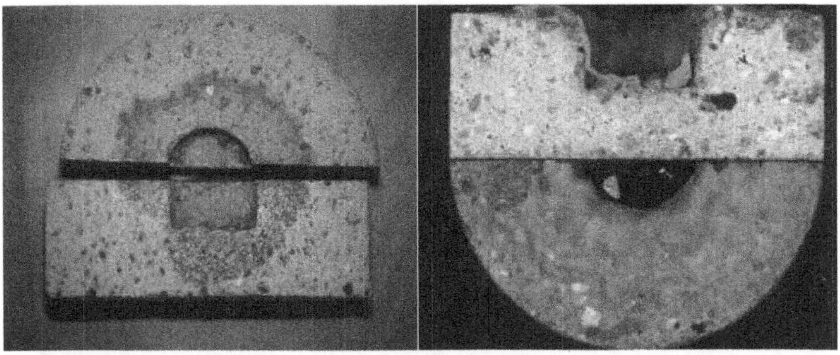

Figure 15. Severe Soda Penetration (Left) vs. Negligible Soda Penetration (Right)

Consequently, a low permeability in conjunction with a glass phase presence are beneficial characteristics for a tin bath block to have and therefore a conundrum is created, low glassy phase content avoids 7" splitting, however insufficient glassy phase could result in nepheline formation as permeability increases.

Air permeability and pore size are interdependent and have an inverse relationship to hydrogen diffusivity in other words:

Low permeability, small pore size = High diffusivity and thermal transpiration

High permability, large pore size = Low diffusivity and potential for flaking

In addition, glass phase presence is important such that Na_2O can be absorbed into an amorphous non- expansive composition. Thus there is a compromise to be achieved not only for permeability but also glass phase; another challenge for the refractory manufacturer.

PRESENT DAYS, NEW MATERIALS

To date there has been a considerable amount of investigation and subsequent development of tin bath block materials which continues to the present. Change can be evolutionary or indeed revolutionary; the latest developments are possibly in the latter category as calcium aluminate is introduced into tin bath block chemistry. The presence of Ca^{2+} in conjunction with low permeability can reduce the rate of nepheline formation to a point at which it is not a problem over the campaign of operation. Alternatively blocks entirely comprised of calcium aluminate provide the possibility of total removal of the detrimental alkali/alumina-silicate interactions whilst providing the thermal insulation that the bath requires. Table 2 below shows the various material types; the criteria established for bath blocks over the years and whether these are still applied or are relevant.

Table 2. Selection of Tin Bath Materials and the Associated Application Criteria (GlasRef Consulting)

	Golden Rules	Ca-aluminate bonded	Hibonite	Ca-aluminate	Hibonite for the Tin Bath Roof
Al2O3 (%)	39-41	40.9	79.3	66-84*	90.0
CaO (%)	N/A	1.4	11..7	10 - 24	9.0
SiO2 (%)	N/A	54.4	6.3	4 - 6	0.84
Major Calcium Oxide Containing Phase/s	N/A	Ca-alumino-silicate	Calcium Hexaluminate. Hibonite	Calcium Dialuminate; calcium aluminate +others	Calcium Hexaluminate. Hibonite
Hydration potential	No established rule	None	None	Possible	None
H2 Diffusivity (mm H2O)	<150	75	<150	<50	Not relevant
Proof Test % Failure Strain	0.4 – 0.7	0.65	TBC	Not known	Not relevant
E ($N/mm^2 \times 10^4$)	>0.5 <1.4	Proof test used in preference.	1.6	~1.6	Not relevant
Glassy Phase (%), nepheline susceptibility	19-26	21 Minimal	Not relevant the material is inherently resistant to alkali	Not relevant the material is inherently resistant to alkali	Not relevant the material is inherently resistant to alkali
Permeability (nPerm)	<2	0.2	0.2 Possibly not of relevance as material is alkali resistant	~8 Possibly not of relevance as material is alkali resistant	0.04

These "new" calcium aluminate materials which are inherently alkali resistance must meet all the previously established criteria to ensure that the problems of the past are not re-visited. As with all panaceas there is a potential drawback to using calcium aluminate in that only one phase, hibonite (calcium hexaluminate) does not hydrate and therefore it is important that calcium aluminate based blocks either do not have hydration potential or have not hydrated prior to installation. To date, to the author's knowledge both the Ca-aluminate bonded and calcium aluminate blocks have not shown nepheline flaking, however it is still relatively early days and only time will confirm whether these new materials can last the 20-40 years campaign duration. An important associated development of the calcium aluminate materials is a new composition for the tin bath roof (TBR) which can be equally affected by alkali vapour transport from the ribbon. Once again to the author's knowledge there has been no report of feldspathoid formation in these materials. SUMMARY AND CONCLUSION

The float process was revolutionary and has subsequently posed unique challenges for the refractory manufacturers. The problems have been unexpected and have taken considerable research backed by laboratory tests to overcome. Fortunately the solutions have been viable and float has grown to the phenomenal success that it is today. The collaboration between refractory manufacturer and glassmaker has been crucial, however the edit remains "follow the golden rules". It is concluded that although it is fervently believed that all bath block issues are known; the history of the float process nevertheless points directly at the "unexpected", nothing should be taken for granted and there are likely to be issues that haven't entered into the arena as yet. Beware.

ACKNOWLEDGMENT

The author would like to acknowledge the immense help and mentoring he received from Mr Geoff Evans (GlasRef Consulting, ex-Pilkingtons); supplying both information for this paper and recounting the tales of bath problems over the years. He was determined that the lessons of the past should not be forgotten and therefore I hope that this paper fulfils this aim. He will be sadly missed. The author would also like to acknowledge Mr John Hartley and Mr Andrew Rushton who were instrumental in the DSF tin bath block development; colleagues at DSF who have contributed and last but not least Sir Alistair Pilkington and his team for their endeavours bringing this truly innovative process to industrialisation and consequently building the glass rich environment in which we exist.

REFERENCES:

1. Bricknell, D.J., FLOAT Pilkingtons' Glass Revolution.
2. Müller, F., Kim, S.K., Gebhardt, F., Küstner, D., Ber.62 (1989) S.369-376 u.418-420

NEW OXY-COMBUSTION CROWN DESIGN FOR EFFICIENT FLUE GAS HEAT RECOVERY

W. Kuhn and A. Reynolds
Fives Stein Limited
United Kingdom

ABSTRACT

Oxy-combustion in glass melting has the advantage to create a reduced amount of flue waste gas compared to aero-combustion and this gives potential benefits of reduced energy consumption and reduced emissions. However, the use of oxygen is expensive and with ever increasing efficiencies of regenerative furnaces the economics of oxy-fuel can be questionable. When oxy-fuel is considered the favorable technology further improvements in energy efficiency are welcome. Even with the reduced flue amount, still some energy is contained in the flue gas leaving the tank at the typical temperatures of 1450°C (in case of soda-lime glass melting) and FSL has targeted this as a potential route to increasing overall efficiency of energy usage. FSL has developed an innovative crown design with a Heat Recover Area (HRA) which encourages optimum heat transfer between the waste gasses leaving the combustion chamber and the raw materials entering the furnace at the batch charging position. The HRA can be realized without an increase in size or footprint and without significant increase in Capex and has been proven to increase efficiencies between 6 and 10% equating to 500kUSD annual savings for a large (300TPD) tank. This paper reviews the history of associated technologies, the basis of the HRA design, how it was derived and finally presents information from the recent installation made by FSL in Europe.

CONVECTIONAL OXY-FUEL DESIGN & HEAT RECOVERY

Oxy-fuel provides an energy efficient and low emission combustion technology however the economics do not always add up when accounting for the higher levels of Opex. In a conventional 300TPD oxy-gas furnace some 40% of energy input does not go into melting glass, and out of this a significant portion is energy lost in the waste gasses leaving the combustion chamber. It is therefore obvious to look at ways to recover some of this heat. Heat recovery by thermal exchange between waste gas and batch is one option and is not new. Special heat recovery tank sections have been proposed in the past to recover energy by exchange with the batch blanket. The drawback of such zones is often increased tank length with additional wall losses that quickly off sets any advantages.

It is crucial for the economic viability of a heat recovery tank section that a high mean heat flux from the flue gas to the batch blanket is achieved. This heat flux must stem uniquely from the flue gas and not from parasitic radiation from the crown and flames of the tank center. This has been solved by separation walls in the past. However, the addition of a separation wall has only limited benefit. In order to create a sufficient radiative heat flux from the flue gasses, it was also traditionally believed that a larger gas volume is required in order to achieve a good emittance. In Figure 1, the emittances for a layer of oxy-flue gas with increasing thickness is calculated according to different approaches in literature. The curves show slight differences but reveal that a flue gas layer of about 1m height has an emissance <0.5 and is not expected to radiate well onto a batch blanket.

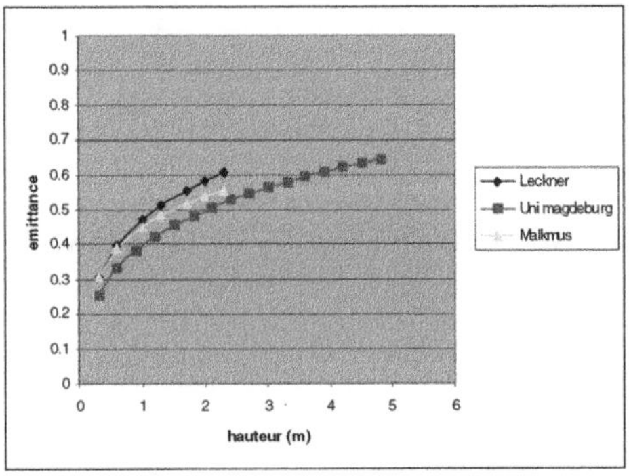

Figure 1. Calculated Emittances for a Layer of Oxy-flue Gas

A NEW APPROACH (THE HEAT RECOVER AREA)

A detailed internal study by FSL revealed that this is not an appropriate approach for the optimization of the heat flux from the flue to the batch blanket. High flue gas volumes above the batch layer favor parasitic radiation and uncontrolled convection that are detrimental to efficient heat exchange. It was discovered that the best arrangement for a high impinging radiative flux to the blanket is a reduced crown height. Reducing crown height at the position of the incoming batch causes the flue gases sweep in a controlled manner over the blanket and limits the effects of parasitic radiation. In a parametric modelling study of different configurations for an oxy tank, the total heat transfer of the flue gas to the batch blanket in a recovery section is monitored. Figure 2 shows the result for different crown heights in the recovery section. The best heat transfer is found for crown heights of about 0.4 m.

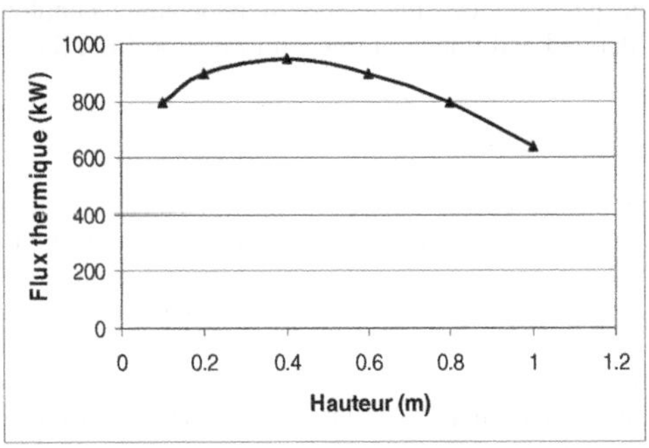

Figure 2. Crown Height vs. Flux

In an actual tank design, it is necessary to find the best compromise between practical feasibility and theoretical optimum configuration. Further practical criteria concern optimizing the velocity and flow pattern of the waste gas in the heat recover section (HRA) to avoid carry over and to avoid any a cold stagnation layer from the batch CO_2 that would reabsorb part of the radiation before reaching the batch surface. By reducing the chamber height, the optical thickness of the flue gas is reduced and reabsorption of radiation is reduced. One has also to bear in mind that the crown participates to the impinging radiation (in absorption bands and in optical windows) and thus upwards directed radiation from the flue gas is recovered. All these factors contribute to high impinging heat flux on the batch blanket even from a quite 'thin' flue gas layer. The reduced crown height has also a series of further advantages:

- Suppression of the thermo-convection roll of the flue gas – always present in traditional combustion chamber designs
- Rapid evacuation of batch gases (CO_2 and else) at lower temperatures
- Reduction of required refractory material

Based on these positive results, an extensive study was conducted by FSL in order to define the best implementation of the new crown design in oxy-combustion tanks of different sizes (Figure 3). Constructive solutions were found starting from small tanks of a few tens of TPD up to float tank size. CFD studies revealed that efficient recovery of the flue gases with savings of 6% could already be achieved without increasing the tank length! For a slight increase of the tank length savings of up to 10% can be achieved. Of course, the ratio of additional structural losses and gains from the flue gas heat recovery has to be studied closely, the balance depends on parameters such as flue quantity, batch coverage, emissivity, tank size, etc. The achievable reduction of the flue gas temperature is also limited by condensation of volatile species in the flue gas and thus dependent on glass type.

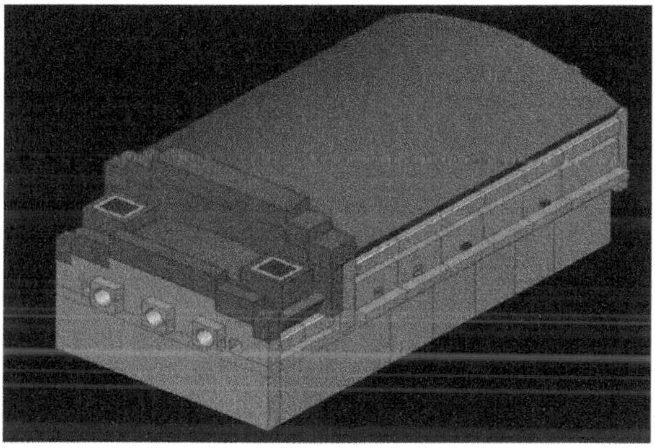

Figure 3. New Crown Design in Oxy-Combustion Tanks

PRACTICAL APPLICATION

FSL has recently constructed and commissioned an oxy-fuel 55TPD combustion tank for borosilicate glass with 2.8MW of installed combustion power (and 2.0MW installed electrical boosting power). The furnace has been designed with an HRA with two flue outlets which has benefits for cleaning and maintenance operations of the ducts in high boron containing glasses. The design optimization of the combustion and the melt space was done by extensive CFD modelling. The numerical modelling of the flue gas exhaust temperature predicted a final exit temperature (at the ports of the HRA) of 1250°C. This has to be seen in the framework of a maximum crown temperature of 1580°C and a mean waste gas temperature of 1560°C in the combustion chamber of the tank. Compared to standard exhaust temperature of this type of tank of about 1500°C, the flue temperature is reduced for 250°C. This corresponds to fuel and oxygen savings of 6-8%.

The tank is now in operation and the measured flue gas temperatures of 1260°C confirm the predicted values! This innovative technical solution has been obtained without an increase in tank size. Further features in the tank design contribute to an efficient melting of the batch blanket and good fining of the glass melt. After almost one year of operation, there is still the need to optimize some aspects of the system; however, the tank supplies a good glass quality with low bubble/seed and stone count under reduced consumption.

CONCLUSION

The HRA gives a practical and economic solution achieving heat recovery from oxy-fuel combustion furnaces, and whilst potential savings are modest in percentage terms, these are achieved without a significant increase in capex of the installation. If we extrapolate the saving achieved in a 55TPD unit to a larger 300TPD furnace these could equate to 500 kUSD per annum therefore making another step toward more economic oxy-fuel combustion.

EXTENDING THE LIFE OF FUSED CAST CERAMICS

R. McGrath and J. Crowe
Johnson Matthey PLC
Royston, Hertfordshire
United Kingdom

ABSTRACT

Johnson Matthey PLC is a company that has been involved with precious metals since 1817, accumulating a vast amount of knowledge and experience. Since the 1990's development of a precious metal coating technology (ACT^{TM} coatings) has been on-going. In recent years the research focus has shifted from the consumable refractories (o-rings, tube stirrers, etc.) to fuse cast AZS furnace blocks. This has been driven by glass manufacturers looking for a solution to excessive refractory wear. This wear impacts both furnace life and glass quality. Until recently the coating could only be applied to machined fused cast AZS surfaces. This paper details the latest innovation of applying the coating to the "as-cast" surface of fused cast AZS blocks and includes micro-graphical analysis and comparison of "as-cast" and "machined" AZS surfaces, details of the latest innovations in the laser pre-treatment process, and results of a three-year plant trial.

INTRODUCTION

Glass is manufactured into a wide range of diverse product from simple containment vessels such as bottles and jars through chemical and industrial glassware, flat glass and high quality lensing to continuous glass fibre for digital data transfer. To everyday users, glass is a relatively benign material but during its manufacture the constituent components of glass require melting at a very high temperature and, in its molten state, the glass is very aggressive towards any material it contacts. Traditionally, alumino-silica based ceramics are used to contain, handle and process molten glass but with the glass itself being a ceramic of similar composition there is inevitable interaction with the containment materials which can give rise to compositional hence property change and incorporation of ceramic 'stones' into the glass stream. Each interaction has consequences on the quality of the glass and the defect rate. While conventional corrosion resistant metals such as steels and nickel based superalloys might be expected to provide some protection in fact the manufacturing conditions are too severe for them to offer any benefit. In practice the only materials able to resist attack by molten glasses have been the platinum group metals (PGM's) and, although the ceramic industries have progressively developed ceramics with improved performance characteristics, their inherent chemistry creates critical limitations. For long service life and for high quality glass the PGM's and in particular platinum and alloys are essential with their unique combination of high melting point and excellent corrosion resistance.

Traditionally Pt alloys have been used in bulk fabricated form or as a cladding over the surface of a ceramic or refractory metal substrate. Pt itself is a relatively soft, low strength material which limits it applicability for load bearing uses and alloying with Rh in particular has long been used to provide the necessary strength and stiffness and hence extend usable surface life. An alternative method of metallurgical strengthening ie. stabilising the Pt grain structure with a very fine, uniform dispersions of oxide particulates to limit grain growth offers even better performance. Johnson Matthey was the first PGM manufacturer to develop a successful grain stabilised Pt, namely ZGS (Zirconia Grain Stabilised) Pt which soon became the market leader; a position that by and large it still retains.

By using Pt alloys and particularly high strength ZGS Pt alloys it is eminently possible to fabricate structural components for processing molten glass. However, the capital cost of these components is high because of the thickness, hence weight of material that is required. For claddings the PGM usage can be significantly reduced by the use of sheet at a lower thickness as there is no mechanical load bearing requirement but there is a practical limit beyond which fabrication of the cladding becomes difficult and costs escalate. Additionally, components of a complex shape can prove extremely awkward, if not impossible to successfully clad.

In either event, the use of PGM's has not been economically viable for high volume low cost glass product such as bottles or tableware. But by the 1990's it was apparent to Johnson Matthey that a better way to use Pt was available; that is to use it only where it's unique combination of properties could be balanced with the inherent robustness of the ceramic substrate as a true composite rather than a disparate multi-element component. By using a thin layer of Pt integrally bonded to the surface of the ceramic the best characteristics of each could be utilised together as a single component while costs could be maintained at reasonable levels. With this balance of cost and functionality the quality benefits imparted by glass contacting inert surfaces could be realised.

COATING DEVLOPMENT

In the middle of the decade, Johnson Matthey launched ACT™ coatings; the first commercial application of bulk PGM coatings for the glass industry. The ACT™ coatings (predominantly but not exclusively Pt) are deposited by a flame or plasma spray process at nominally room temperature and typical coating thicknesses are in the range 200 – 300 µm which equated to ~40% of the cladding thickness at the time of launch although this has subsequently been reduced to better compete with ACT™. Initially the new ACT™ coatings were targeted towards the end of the glass production line ie. the forehearth, with thermocouple sheaths being the lead product closely followed by forehearth consumables such as plungers, tube stirrers and orifice rings (Figure 1). Once ACT™ coatings achieved market acceptance the intention was to move their usage back 'upstream' towards the furnace itself where operating conditions were more demanding and the

Figure 1. Selection of Typical Orifice Rings

required service life was longer. Initial developments of ACT™ coated lip blocks and tweels were successful but difficulties were experienced with the dense fusion cast AZS (alumina-zirconia-silica) ceramics used for critical components with the furnace itself. The philosophy of ACT™ coating had always been to develop and use the asperities typical in the surface of slip cast ceramics as a key to coating adherence but the high density and toughness of the fusion cast materials meant that the fine porosity necessary for keying were not present. Further, mechanical methods of roughening the surface proved ineffective.

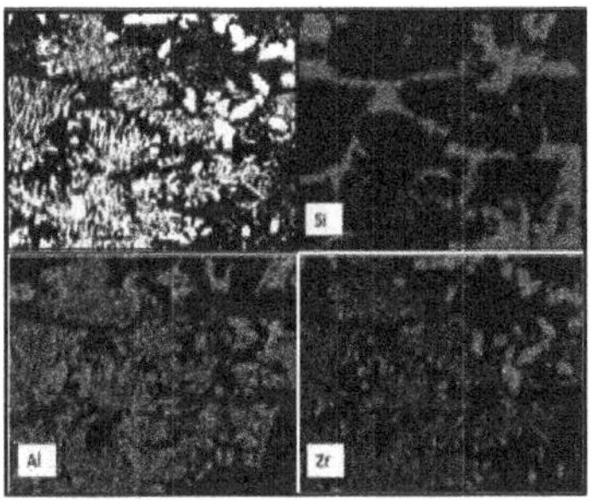

Figure 2. SEM Energy Dispersive X-ray Spectrometry Maps

As might be expected by the method of manufacture, the inherent structure of AZS ceramics mimics that of metals. Whereas the overall elemental content complies with the stated composition, on a micro-level the ceramic is effectively inhomogeneous. It is a seemingly random mix of silica, alumina and zirconia rich 'phases' with a varied microstructural directionality related to cooling kinetics, the crystallisation of the differing constituent phases and the variable cooling rate from centre to surface of the cast form, which generates the distinct phases and orientations. The compositional variation is clearly shown by the image of x-ray mapping of a section from an AZS ceramic block (Figure 2). Additionally, the cast ceramic develops a 'chill cast' surface skin during solidification consisting of a much finer phase distribution, rich in alumina and zirconia which is significantly more resistant to mechanical roughening. This is illustrated in Figure 3. From the viewpoint of applying an ACT™ coating, the result is that while mechanical roughening might be feasible in certain areas of the ceramic surface it may be less effective in others. Attempts to uniformly roughen the complete surface proved unreliable. Both coarse and fine media of several compositions was evaluated using venturi and pressure pot blast systems but the resulting surface developed an uneven macro rather than micro roughness and as such the adherence of any deposited ACT™ coating tended to be variable.

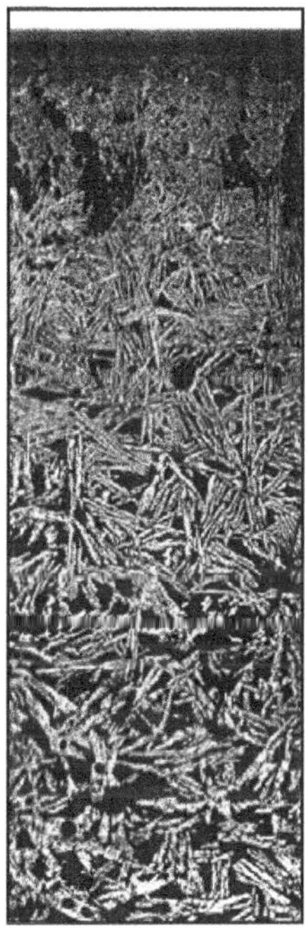

Figure 3. Section through Fusion Cast AZS Block Showing Finer Phase

LASER DRILLING DEVELOPMENT

Surface preparation of AZS ceramics was therefore a major issue to the effective deposition of an ACT™ coating. As a first step Johnson Matthey insisted that the cast surface of any such ceramic

This was not particularly restrictive as fusion cast AZS is manufactured with the need to machine to finish form in mind; if not for simple quadrilateral furnace blocks then for a more geometrically complex outlet or lip blocks. Chemical etching of the silica rich constituent phase with an HF containing solution proved successful on trial samples but the variable orientation of the AZS ceramic microstructure still created difficulties which meant that it wasn't effective in all areas. Additionally, the H&S risk associated with bulk usage of HF required dedicated facilities and close monitoring of operation and, on consideration, the etching approach was rejected.

A third option considered was the use of a bespoke interlayer via. the application of a ceramic wash-coat. A great deal of effort went into the identification of a potential coating systems which, while initially showing promise, was once again defeated by the microstructural variability. A further complication was added by the invariable need to fire the wash-coat to bond it to the AZS substrate. In general, the bond strength of any wash-coat formulation improves as the firing temperature is increased but with AZS ceramics being susceptible to thermal shocking, heating and cooling rates were very slow at $<10^{o}C$ per hour which meant that a heat treatment cycle extended over many days and the risk of catastrophic failure was ever present with no knowledge of the outcome until the furnace was opened. The large size and weight of the blocks needed by the glass industry also required specialist furnaces and pushed the capacity of available facilities.

A potential solution to the problem of improved and consistent ACT™ coating adherence came with the realisation that laser surface restructuring being promoted in several technical publications in the mid 2000's could be applied to ceramic materials to develop an appropriate surface 'key'. Johnson Matthey were already interested in lasers for precision cutting of high melting point PGM's and it was believed that the use of this equipment could be extended into the area of pre-preparation for ACT™ coating.

Options were explored with different laser manufacturers and different beam generation systems. Samples of typical AZS and slip cast consumable ceramics were laser drilled with different lensing over a range of operational parameters to produce a matrix of 'pegs' (examples shown in Figures 4 and 5) after which the treated surfaces were ACT™ coated with Pt. Comparing the coating adherence with the 'peg' profile enabled the optimum parameters to be identified; a reasonable degree of flexibility being required to accommodate the same effects of varying microstructural composition and orientation identified earlier. A further level of complexity was introduced by varying the 'peg' density by altering pitch and spacing of the drilled holes and ultimately parameter sets were identified for the standard fusion cast AZS ceramics and the standard ACT™ coating materials. While it might be assumed that a higher the 'peg' density was preferable, giving more asperities for the deposited Pt coating to mechanically grip, this had to be balanced against a theoretical and practical weakening of the ceramic surface. Ultimately the correct combination was identified and a typical laser drilled surface is shown on the left below.

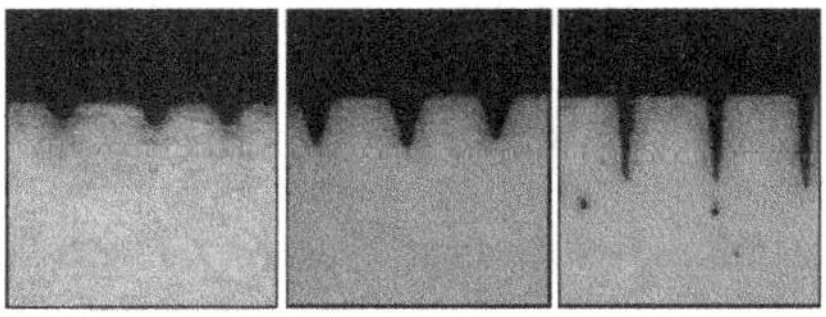

Figure 4. Blind Holes Laser Drilled into the AZS Ceramic Test Blocks

Figure 5. Typical Laser Drilled Fusion Cast AZS

With the desire to consolidate the laser drilling into the Plant already used to deposit ACT™ coatings the focus was initially on pulsed Nd-YAG lasers. However, the excellent beam density of continuous wave Yb fiber lasers were identified as an alternative that better fitted Johnson Matthey's other interests in laser technology. Both systems are fully capable of drilling the range of ceramic materials traditionally coated for the glass industry as illustrated in Figure 6, which shows an ACT™ Pt coated laser drilled AZS test block.

Figure 6. Pt Coated Laser Drilled AZS

As might be expected there are downsides to having to prepare AZS ceramic blocks by laser drilling. An obvious one is the time involved in completing this extra processing stage. Although the laser drilling is completed 'on the fly' rather than in a stop/start sequence it is effectively a duplication of the final ACT™ coating process. Further, the geometry of the substrate can present difficulties with respect to the positioning and required movement path of the laser head on the robot arm. Complex curves can necessitate dividing what would be a single movement into a number of smaller sequences and take more time to programme. Cylindrical bores through the substrate or relatively narrow channels in the surface can also present problems if the access is insufficient to allow access to or free movement of the laser head. Johnson Matthey therefore has to put restrictions on what can be achieved, however, with consideration and some design modification most component shapes can be accommodated. The fact that the surface of the AZS component needs to be covered with a multitude of 'pegs' also means that the final weight of the ACT™ coating is proportionally increased by the metal required to fill the drilled 'pegs'. Typically, this adds 10 - 12% to the weight of metal required.

To ensure complete coverage of the ACT™ coating Johnson Matthey strongly recommends that the coating thickness is sufficient to completely cover the laser drilled surface to give a reasonably smooth even finish. Holes, typically open porosity, in a substrate surface can fill unevenly due to shadowing effects (one side of the hole masking the line of sight coating deposition) and if a free path is left to the underlying ceramic the triple point juncture between Pt coating, ceramic substrate and molten glass can create an electric cell and the subsequent gas evolution which generates bubbling. The profile of the laser drilled 'pegs' is such that shadowing is minimised but, typically, a minimum ACT™ coating thickness of 300 µm would be recommended with 400 µm preferred in critical glass contact areas.

With respect to surface porosity and the AZS itself, reference should also be made to the expected condition of AZS substrates supplied to Johnson Matthey for ACT™ coating. As a fusion cast material the presence of random porosity on the machined surfaces of the AZS is common. A proportion of this porosity might more correctly be defined as a 'blow hole' as the cross section can be more than 10 mm wide. If excessive, Johnson Matthey reserve the right to reject the component but within an acceptable range the porosity and holes can be filed with a ceramic cement. A room temperature setting formulation which laser drills acceptably has been identified but excessive use can compromise coating adherence if the locally filled area is large.

In a recent example a ceramic was supplied to Johnson Matthey with a complex curved geometry. This geometry was such that the surface could not be machined back the required 25 mm. Therefore, laser drilling had to be attempted onto the un-machined as cast surface. The images in Figure 7 depict the surface before and after coating. Prior to coating a number of trials were run on small 'as cast' samples to ascertain the best laser drilling parameters. After a number of experiments final parameters were selected that were slightly different to those used for machined surfaces. These blocks were installed in 2014 and are still in service today.

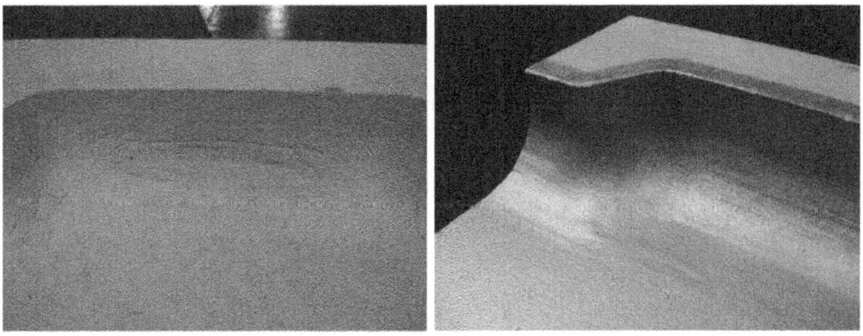

Figure 7. As-received Fusion Cast AZS block (Left); Same Block After Laser Drilling and ACT™ coating (Right)

To evidence the performance potential of ACT™ coatings on fusion cast ceramics this year, a furnace containing an ACT™ coated AZS block was rebuilt. The block was returned to Johnson Matthey for recycling. The surface of the block remained in very good condition. To confirm the condition of the coating samples were taken and analysed. The representative image in Figure 8 shows the ceramic, ACT™ coating and glass. Also visible is a laser drilled hole filled with platinum. The nominal thickness of the ACT™ coating was 300 μm when shipped and measurement confirmed that this has been maintained throughout the 4 years life of the furnace.

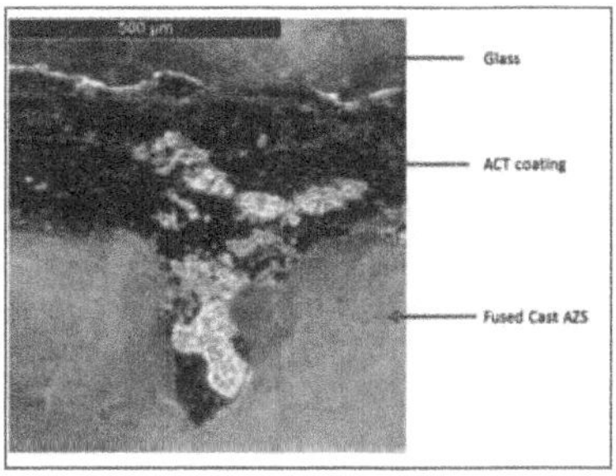

Figure 8. AZS Block Retained Integrity of the Pinning Methodology

CONCLUSION

In conclusion the ACT™ coating process has developed considerably over the past two decades. In particular the laser drilling pre-treatment signified a real step change in capability. Now in use in numerous campaigns of up to eight years ACT™ coating can be considered a genuine solution for fused cast AZS block protection as well as less arduous thermocouple and consumable applications.

ENHANCED RADAR CONTROL FOR HIGH PERFORMANCE BOTTOM PAVING

J-G. Vuillermet, M. Gaubil, I. Cabodi, and O. Bories
Saint-Gobain CREE
Cavaillon
France

ABSTRACT

Based on SEFPRO R&D knowledge on microwave technology to analyze internal structure of AZS fused cast soldier block we developed new tools to be able to prevent blind area obtained with conventional radar wave's analysis on small thickness tiles. The advantages of this new technics is to obtain reliable internal structure pictures of AZS fused cast tiles used in bottom paving application for containers of float glass furnace. The strength of this microwave technology allow to guaranty the high corrosion resistance of our fused cast tiles. Combined with tight joint machining and the double layer refractory mortar solution, it's constitute a reliable high performance bottom solution well adapted to specific conditions such as high insulation to prevent heat loss, electrical boosting or extra white glass melting.

INTRODUCTION

The quality of a pavement is critical for glassmakers as glass leakage is of course a major issue. But some trickier problems specific to bottom area – such as downward penetration of metallic elements in the sub-layers, upward drilling in tiles – are also very difficult to handle. Especially for furnaces operating in harsh condition, namely high temperature, highly convective batch, frequently changing conditions of operation, it will be important to secure a reliable bottom assembly lasting life. As a first step, the use of fused cast AZS, with different level of zirconia and thicknesses will be consistent with this objective , with high corrosion resistance to sodalime glass. Nevertheless, for as cast AZS tiles, internal filing is key in terms of corrosion resistance, Figure 1 shows that a dense material has higher corrosion performances.

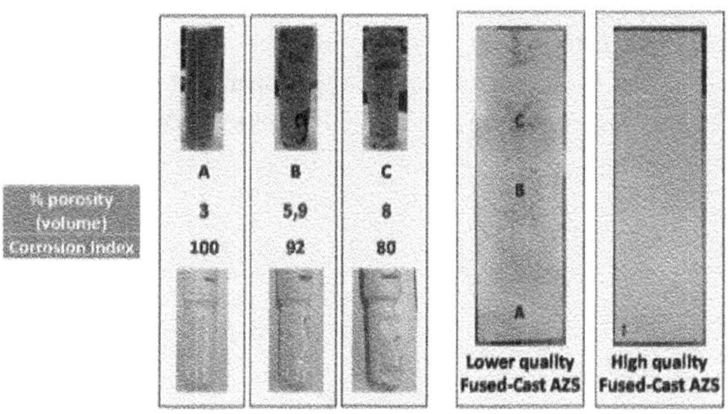

Figure 1. Corrosion Resistance of Dense vs. Porous Areas in a Fused-Cast AZS Tile

The presence of internal porosity can decrease by 20%, corrosion index of the fused cast product. The internal filling control of the tiles becomes necessary and this is not achievable with the standard radar solution.

ENHANCED RADAR SOLUTION FOR THE INTERNAL FILLING CONTROL OF LOW-THICKNESS PRODUCTS

BACKGROUND: THE GROUND PENETRATING RADAR

Originally, radar detection has been developed for soil investigation with for instance archeological, geophysical or civil engineering purposes (SAGNARD & REJIBA, 2016). Electromagnetic waves in the frequency range of 10MHz to 2 GHz are emitted by the device as pulsed waves (see Figure 2 for explanations). Translating the device upon the area enables a comprehensive inspection. At such frequencies, a penetration depth of 0.1m to 100m can be achieved in classical soil materials. It is important to note that, the higher the operating frequency, the lower the depth, but the better the resolution.

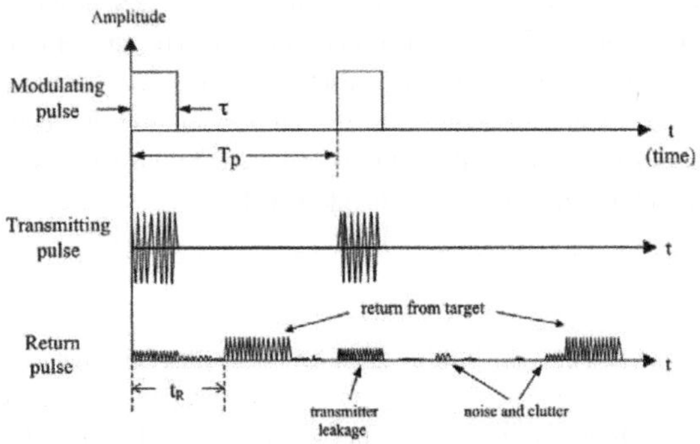

Figure 2. Modulated Pulsed-Wave Emission in the GPR Applications

IMPLEMENTATION FOR REFRACTORY PRODUCTS

A refractory block is definitely a very thin soil with higher attenuation properties in respect with an earth soil. The typical size of defaults that are contemplated is about 5cm or less. So, as discussed above, radar detection in refractory will be achieved with as high operating frequency as possible. For blocks inspection, the nominal operating frequency of the device is 1.6 GHz, which is in the very high range for classical GPR devices (BOINET, et al., 2011).

The so-called radargrams that are derived from the inspection have to be processed. Indeed, the resulting signal's interpretation is not straightforward. A spot default in the structure will generate a hyperbolic shape in the radargram (see Figure 3). Some automatic post-processing steps were developed by SEFPro with dedicated software to generate a more user-friendly picture of the inner structure of our soldier blocks, as can be seen on Figure 4.

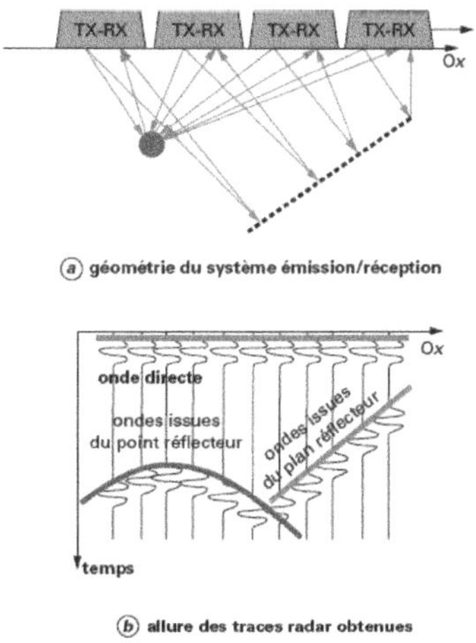

Figure 3. Basics of the GPR - a spot in the soil (grey, above) will generate a hyperbolic shape (in red, below) in the radargram

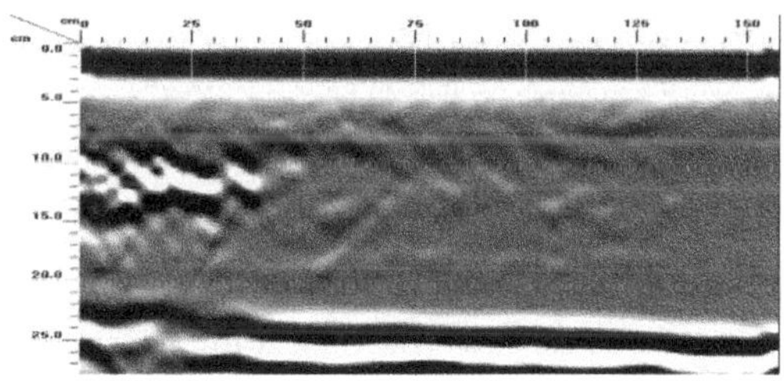

Figure 4. Typical Radargram Obtained after the Hyperbolas-Removal Step

WHAT ABOUT THIN STRUCTURES?

Nevertheless, as we can observe on the Figure 4, the two interfaces between refractory and environment, lead to undefined stripes between 3 and 5 centimeters in thickness, that can be associated to two different effects:

- On the upper interface, it corresponds to a systematic blind zone because part of the energy emitted in the pulse is directly transmitted to the receiver
- On the upper and bottom interface, the interface definition is managed with a broad uncertainty by the device called border effect. This discontinuity between two media will send back an echo during a time span of τ that generates poorly defined areas which screen any other echo.

Practically, these effects induce on the radargram two strips around 3 to 5cm broad. For soldier blocks with 250 mm thickness, this will not be an issue, as the critical defects are mostly localized within the core (namely the shrinkage). But for tiles, this appears to be a problem, as 50 to 100% of a radargram will be covered by those meaningless effects.

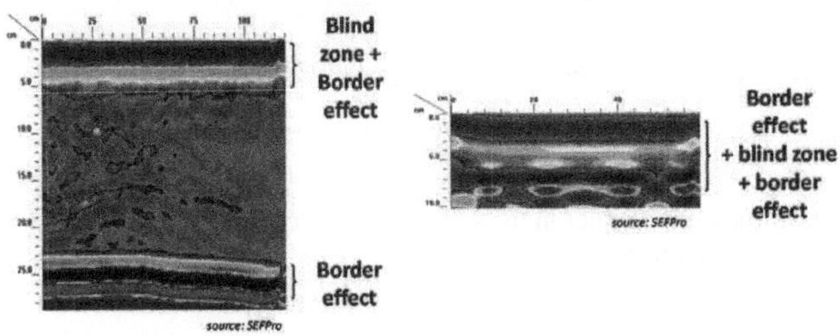

Figure 5. Problem of Thin Structures with Classical Radar Device (relevant inspected area is narrow)

HOW TO SOLVE THE PROBLEM

As discussed, the pulse duration τ will be the key parameter to play with in order to decrease the blind zone and the border effect (BORIES, 2016). To achieve this reduction without losing Signal / Noise ratio, some heavier systems had to be implemented. With this new device more than 60% thickness in the tiles is available for relevant radar inspection.

For the soldier blocks, a manual inspection can be easily performed which is not the case with the new, more cumbersome device on tiles. As a consequence, radar control was implemented directly on the line, with several scanning paths on the tiles. With this new radar system, a systematic analysis is done on the production line, leading to relevant quality control of the products.

ENHANCED BOTTOM SOLUTIONS: WHERE WE ARE

Assessing the good filling of the bulk fused cast AZS material is necessary. But this is only part of the overall paving performance in preventing from all type of potential damages that can affect a furnace's bottom. It needs to combine with two key features of a bottom's assembly, that are discussed in the following part.

JOINTS TIGHTNESS TO AVOID GLASS PENETRATION

Having tied joints is a prerequisite to avoid glass penetration towards the bottom sub-layers. Recent advances in dimensional control technologies enabled to assess the relevance of SEFPro's TJ specification (joint opening <0.5mm). A *blind test* was performed on two standard tiles (not TJ, Figure 6), then on a four TJ tiles assembly as shown on Figure 7Figure 6. The assembly was controlled numerically on the whole surface of the contact faces. This extremely precise kind of control cannot be achieved on a whole pre-assembly as the tiles are not digitalized on a 100% basis. But thanks to both structured light projection and mechanical palpation tools, we could certify that our sample complied with specifications.

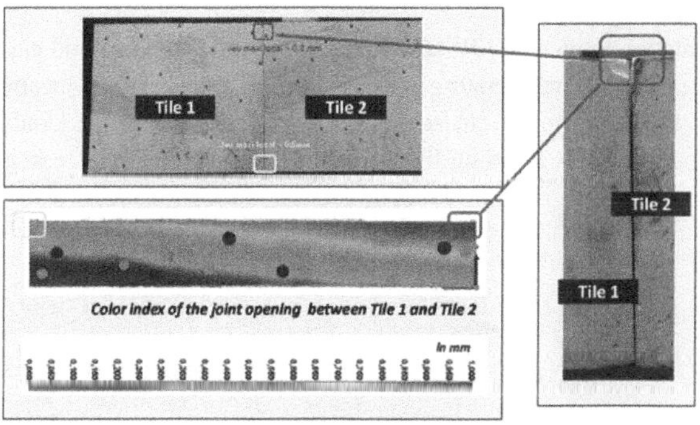

Figure 6. Standard Joint Assessment with Digitalization Tools - Two Tiles

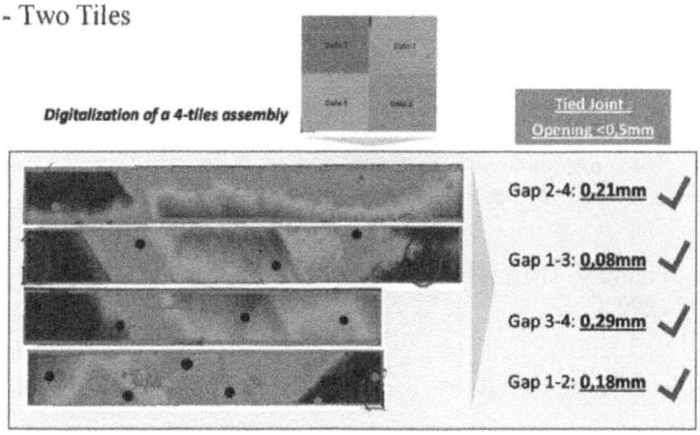

Figure 7.TJ Tiles Joint Assessment with Digitalization Tools - Four Tiles

MULTI-LAYER SOLUTION

In order to guarantee to ever tougher operating glass furnaces a secure sub-layer, a reliable composite solution (see Figure 8) had to be developed: this is essentially made of fused-cast AZS grain refractory mortar, with good corrosion resistance in contact with soda lime glass.

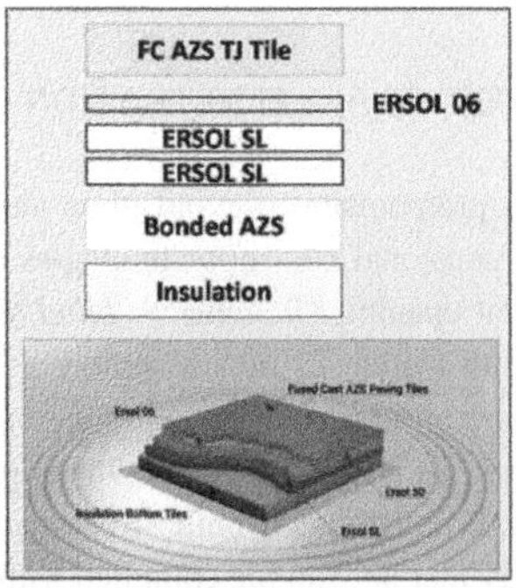

Figure 8. Multi-Layer Solution for Furnaces Bottom

A new formulation known as ERSOL SL (Self Levelling product and easy to handle) is well adapted to high bottom temperature because during the furnace's heating-up, thermal expansion behavior keeps the joints closed until 1250°C (see Figure 9). Doubling this layer allows refractory mortar to show a sliding-like behavior between each other, resulting in a crack-free sub-bottom assembly.

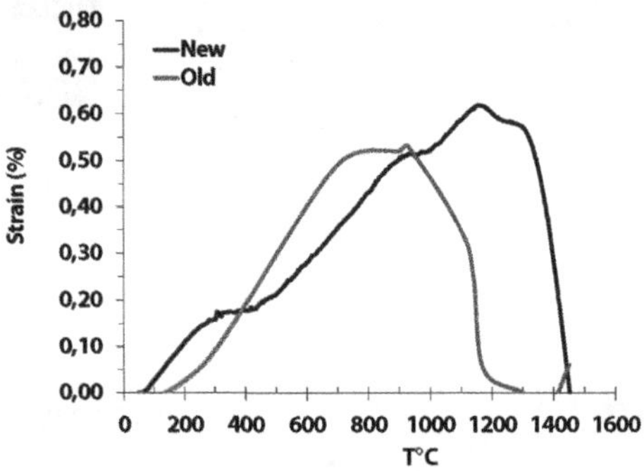

Figure 9. Thermal expansion of ERSOL SL at 1200°C

CONCLUSION

Furnace bottom is a complex thermo-mechanical system, subject to aggressive interactions with the soda lime molten glass. In order to maximize the bottom assembly performances, new radar analysis technics have been developed to get right internal filling pictures on fused cast AZS low-thickness tiles. Based on digital controls made on randomly chosen samples, we demonstrate the actual relevance of TJ specifications. Precisely machined (TJ specification) FC tiles, insuring tight assembly, combined with the double layer of fused cast AZS grains based refractory mortar offers maximum reliability to bottom solutions.

REFERENCES

BOINET, M., GAUBIL, M., PIRES FRANCO, P., CROUZET, A., MASSARD, L., & NORGEOT, C. (2011). Localisation of structural heterogeneity in refractory blocks using ground penetrating radar. *UNITECR.*

BORIES, O. (2016, March). *Patent No. WO 2016041902 A1.* World.

SAGNARD, F., & REJIBA, F. (2016). Géoradar - Principes et applications. (E. T.I, Ed.) *Techniques de l'ingénieur - Applications radars, base documentaire : TIB592DUO.*

HIGH PERFORMANCE SUPERSTRUCTURE CONCEPT FOR CONTAINER GLASS FURNACES

B. C. Snow and T. Close
HarbisonWalker International
Pittsburgh, PA

ABSTRACT

Superstructures in container glass furnaces are critical to the performance of the melter. The key performance indicators of a melter influenced by a superstructure are: Glass quality, Furnace life, and Energy, efficiency, and emissions. Using a combination of some available technologies will enable furnace designers to address all of the key performance indicators of the melter. When compared to traditional technologies, high quality sintered AZS materials provide engineering alternatives that result in more energy-efficient melters and improved glass quality while maintaining the desired campaign life. High emissivity coatings enable heating of a melter to operate more efficiently, lower emissions through the stack and improve alkali resistance of the hot-face refractories. Combination of sintered AZS materials, high emissivity coatings and strategic design of the refractory lining will impact all of the key performances indicators making a high performance superstructure possible.

INTRODUCTION

In most glass making furnaces, where high quality glass products and long campaign life are concerned, the glass contact zones will typically get the most attention. However, superstructures in fuel-fired, soda-lime glass container furnaces also play a critical role and attribute greatly to the furnace energy efficiency. In the past decade, the focus on energy efficiency has dramatically increased on all energy intensive processes; including glass making. New technologies, like sintered AZS materials and high emissivity coatings enable furnace engineers to achieve long campaign life, improve glass quality and increase energy efficiency.

ENHANCING PERFORMANCE WITH SUPERSTRUCTURE DESIGN CHOICES

The use of fused cast AZS in superstructures in soda-lime glass furnaces has been the standard material for several decades. It well known for resistance to glass corrosion and alkali vapor attack. Unfortunately, this material is also well known for having high thermal conductivity, a thermal expansion hysteresis and a tendency to exude the glassy face out of the refractory body. It is because of the latter attributes that P. Carlo Ratto wrote a paper named, "The end of fused-cast refractories in soda-lime glass furnace superstructures?" In this paper he stated that the "drawbacks of fused-cast can become critical when the performance demand is pushed to the limit" (Ratto, 2013). One scenario outlined where performance demand is pushed to the limit is in oxy-fuel fired furnaces. In this case, sodium vapor concentrations are found up to five-times than typically seen in traditional fuel fired furnaces. When higher concentrations of sodium vapors are present, increased alkali attack is expected. Alkali attack is typically most severe at the joints. One way to combat increased wear of a superstructure lining is to reduce the number of joints.

When designing a superstructure, furnace designers are often limited by the sizes and shapes of refractory products as determined by their manufacturing process. In the case of fused cast refractories, complex shapes are often problematic in the manufacturing process due to the casting method. Alternate technologies, like sintered AZS, have several different methods to manufacture the desired shapes. Pressed brick technology is limited by the size of the presses available. Slip cast technology can definitely handle the complex shapes, however is typically utilized in smaller sizes; especially those with thin walls. Dense-cast technologies are typically available under two platforms: cast, low-moderate heat treatment and cast, high-fired heat treatment. In both cases, these casting processes utilize large custom molds made from various materials like wood, steel, or even polystyrene. The first step of the process is only limited by the production sites lifting capability and the size of the dryers. The two technologies diverge at the post-drying stage. For cast shapes exposed to the low-moderate heat treatment, strength develop relies solely on the chemical binder; the ultimate strength of these materials is achieved in service. For high-fired products, shapes are further exposed to heat in a kiln that can reach temperatures between 2500-3200°F. The bonding strength comes from the sintering process, where atoms in the grains diffuse across the boundaries of the particles to fuse together (without reaching the melting point of the material). In the case of AZS materials, the ceramic bond can only be achieved at high temperatures. The kiln size plays a role in the size limitations, but it can also present some more obstacles. The stresses that the refractory shape may undergo during high-firing, can cause a very large shape to experience some warping or deformation. This can only be corrected by utilizing cutting or grinding machines; these machines have weight capacity limits associated with them as well. Regardless of chemical bond or ceramic bond, once the shape leaves the refractory producer's site it has to be safely transported to the glass maker and installed into the furnace. Lifting capacities during these steps need to be identified when designing a furnace. After everything is said and done, the weight of the refractory block becomes is the most limiting factor. Blocks weighing up to 3500 or even 4000 lbs. (1588 -1814kg) are manageable; this translates to a block that is as big as 82" X 48" X 9" (2083 X 1219 X 229 mm). A block of this size would replace roughly nine fused-cast AZS blocks, which would be a reduction of twelve joints. This is twelve areas of the furnace that are more susceptible to alkali vapor penetration that would be mitigated.

In the redesigned regenerator ports shown in Figures 1 and 2, furnace engineers were able to reduce the number of joints by 60, while still making the shapes in reasonable sizes to aide in the construction of the furnace. Something to keep in mind when designing these large shapes is the handling of the pieces not only during shipment from the manufacturer, but during construction of the components on site. It is important to make refractory shapes that are easily lifted and have symmetry in order to maintain the center of gravity on the item. This is crucial for the safety of the refractory, but also the safety of the installation crew. A side benefit of the redesign is time savings; installing fewer components will take less time. Moving from standard fused cast AZS blocks to larger sintered AZS shapes will have some benefit in decreasing installation time, but moving from standard series brick to larger shapes has substantially more installation time savings.

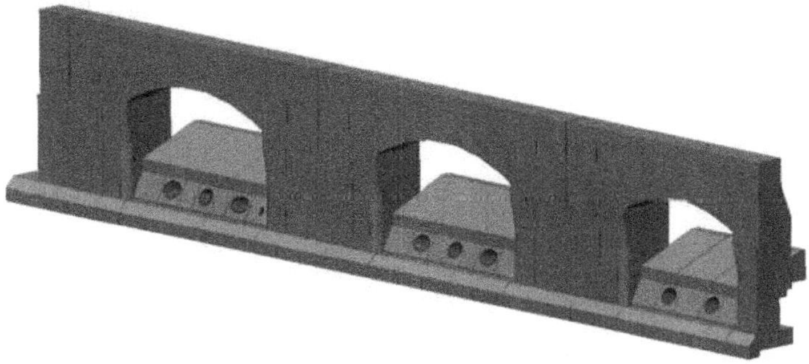

Figure 1. Original Regenerator Port Design with Fused Cast AZS

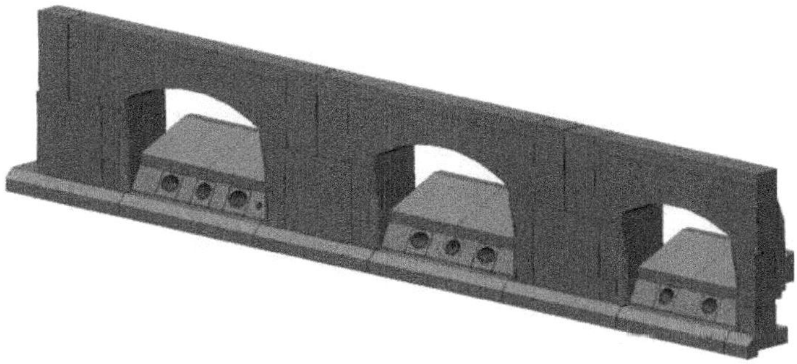

Figure 2. Re-design of Regenerator Port using Sintered AZS

In addition to having greater flexibility in the size of the refractory shape, a sintered AZS material also enables the hot face refractory thickness to be reduced. Fused cast AZS refractory shapes supplied for the superstructure application, are typically manufactured using the regular casting method. This will leave a gate scar on the cold face of the refractory, meaning that there is actually less solid refractory than the lining thickness. With a sintered AZS material, the manufacturing process ensures that the material is solid through the entire shape; this in combination with a lower thermal conductivity allows for a thinner hot face refractory lining. By decreasing the hot face thickness, designers can now increase the back-up insulation layer thickness, which leads to higher energy efficiency. The details around this will be further discussed in a later section.

ENHANCING PERFORMANCE WITH SUPERSTRUCTURE MATERIAL SELECTION

While removal of joints in the design can bring great benefit to the alkali resistance of the superstructure at the macro-scale, the alkali resistance of the refractory itself plays a critical role in the overall performance. Factors that affect the alkali resistance of a refractory include the chemical composition, mineralogical make-up and the pore structure. A study was conducted on four different sintered AZS samples comparing these three factors and how they relate to alkali resistance. The chemical composition of four samples, identified as Sample A, Sample P, Sample V and Sample Z, is shown in Figure 3.

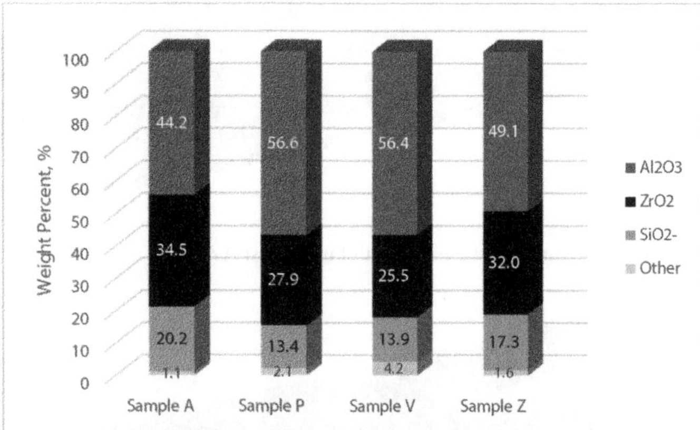

Figure 3. Chemical Analysis by X-Ray Fluorescence (XRF) Method

When the AZS components are stoichiometrically compounded to form mullite and zirconia, the resulting product has both good creep and good alkali resistance. Stoichiometric AZS is: Al_2O_3 = 45.2%, SiO_2 = 17.8%, ZrO_2 = 36%. Sample A is closest to stoichiometric AZS and has no free alumina which should provide this product with the optimum creep resistance and alkali vapor resistance.

As with any ceramic material, the chemical composition only provides part of the story. To gain a full understanding of the refractory system an SEM/EDAX analysis should be conducted to see the microstructure and determine the mineral phases present. See Table 1 for complete summary of results.

The mineralogical make-up of any given product has a direct impact on the performance of the material. Additionally, the matrix and the morphology of such being just as important as the aggregate it is surrounding. The denseness, smooth/rigid appearance, or degree of porosity helps to indicate, on an analytical level what the possible makeup of matrices could be. In the study, four different materials were analyzed through the use of scanning electron microscopy and elemental data was collected by means of energy dispersive spectroscopy in order to confirm a phase's potential identity.

Table 1. SEM/EDAX Data

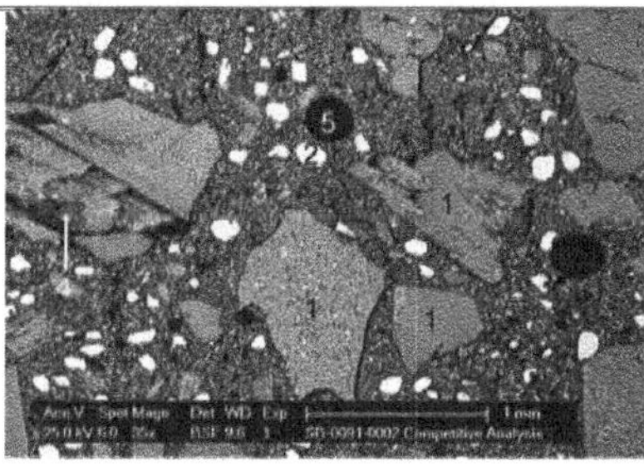 Sample A (35x): 1. Fused zirconia mullite, 2. Zircon, 3. Al-Zr-Si-O matric, 4. Alumina remnant, and 5. void/epoxy	 Sample P (35x): 1. Fused zirconia mullite, 2. White fused alumina, 3. Zirconia, 4. Al-Zr-Si-O matrix, and 5. void/epoxy
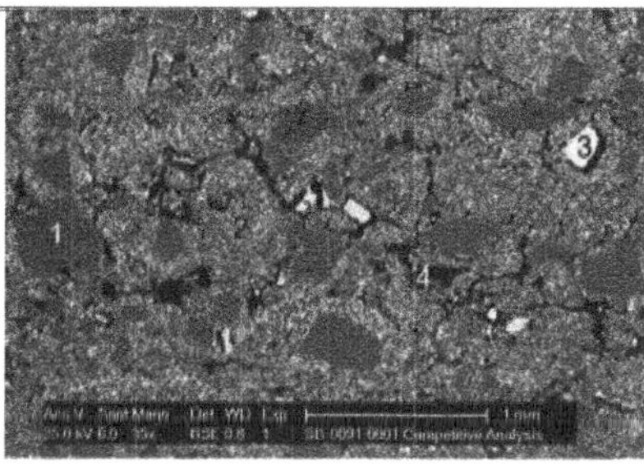 Sample V (35x); 1. Alumina, 2. Al-Zr-Si-O Matrix, 3. Zircon, and 4. void/epoxy	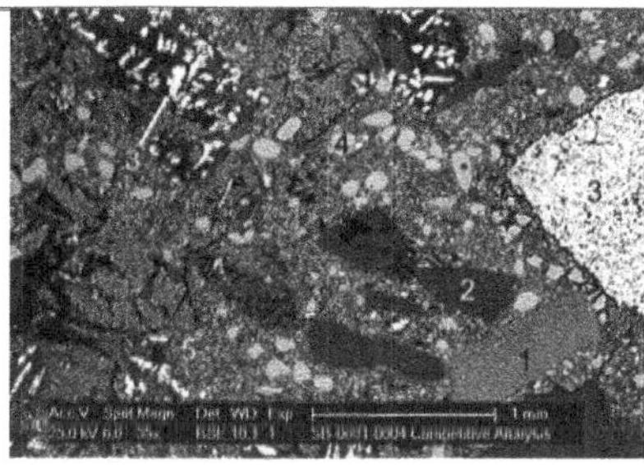 Sample Z (35x): 1. Fused zirconia mullite, 2. Alumina, 3. Zirconia, 4. Zircon, 5. Al-Zr-Si-O Matrix, 6. void/epoxy
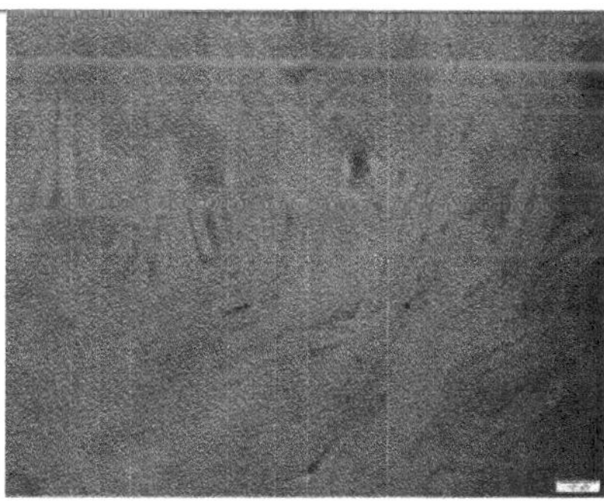 Fused Cast AZS 34% (56x)	

While each sample contains alumina-, zirconia-, and/or silica-based materials in the matrix, the physical morphology and makeup of the finished product differs amongst the four samples. On a microscopic level, visual clues such as the degree of porosity, smoothness/rigidity, or denseness during the mineralogical examination lead to the make-up of a mix. In regards to aggregate, porosity tells a lot about how it was prepared; if there are numerous pores, the inclination is to believe that it was fired or sintered as a raw material, whereas a smoother aggregate surface with low porosity suggests a well-fused material. When considering the matrix porosity, if there are large pores/voids present, it may lead to identify the production method and/or the bonding technology. This difference in porosity between manufacturing technologies are not typically a concern since they do not influence the critical aspect: permeability.

The matrix in Sample A is smooth and continuous – there are no distinct boundaries, but rather one, continuous mass. This morphology suggests that it would likely be an in-situ formation based on the type of brick/material it is. Samples P and Z are noticeably different from that of sample A. The pore distribution, for example, is not as homogeneous and uniform as sample A. The smoothness and continuity of the matrix material also suggested that it was an in-situ matrix formation for this given material. The fact that there are distinct and in-tact aggregate found surrounded by the in-situ matrix formation suggests this as well. Sample V differs from the rest of the samples in that the aggregate is not formed with a fused zirconia mullite grain, but rather alumina and zircon. Lastly, samples A and V both have more matrix volume than samples P and Z. These four samples were tested in a cyclic alkali vapor test. This test requires the samples to be prepared in 3.5" X 3.5" X 1" (88.9 X 88.9 X 25.4 mm) sizes to serve as a lid on an alumina crucible. The crucible holds sodium carbonate, sodium sulfate, sodium hydroxide and water as identified in Table 2.

Table 2. Crucible components in cyclic alkali vapor test

	Cycle 1	Cycle 2
Sodium Carbonate	1.8 g	3.6 g
Sodium Sulfate	1.8 g	3.6 g
Sodium Hydroxide	9.0 g	18.0 g
Water	4.0 g	4.0 g

The sample and the crucible are then brought up to 2500°F (1371°C) for a 24 hour hold in a natural gas fired furnace. After cooling, the sample is removed, placed on a slightly larger crucible with the same contents and brought back up to the same temperature for a second cycle. Upon completion of the test, observations are made with regards to cracking, warping, splitting, peeling, and alkali build-up. Samples are measured before and after testing to determine the linear change of each test specimen. Results of this test, after 2 cycles, on the four samples identified earlier are shown in Table 3.

Table 3. Cyclic Alkali Vapor Test Results

Sample	Expansion, %	Cracking	Peeling	Warping, in.(mm)	Photos
A	<0.5%	none	minor	0.0	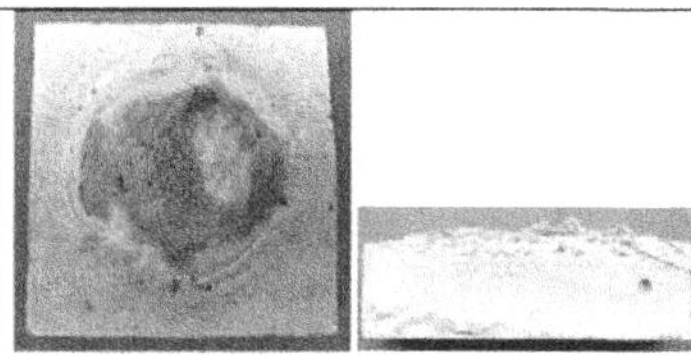
P	4.1%	small open cracks	none	0.04" (1mm)	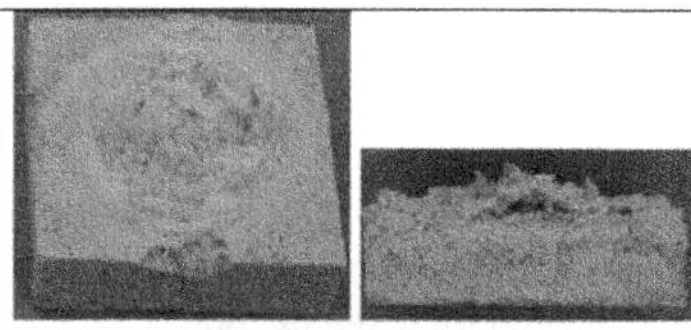
V	0.5%	none	minor	0.0	
Z	3.1%	none	open cracks	0.0	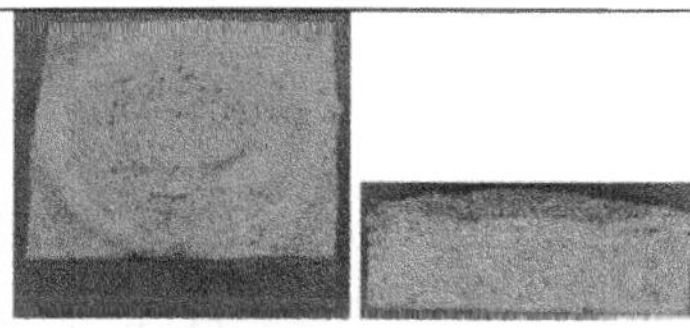
Fused Cast AZS 34%	1.8%*	none	none	0.0	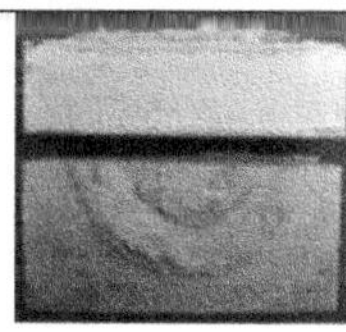

* Expansion caused by exudation of the glassy phase in the sample- not by alkali attack.

None of the samples demonstrated a complete failure of the cyclic alkali vapor test. To further distinguish between the samples and to compare against fused cast AZS, creep resistance and thermal shock resistance was studied. The correlation between the amount of matrix volume seen in the SEM/EDAX images versus the level of alkali attack on the sample could be made. It is hypothesized that alkali attack is related to pore structure of the refractory. In the case of samples A and V, the increased volume of matrix versus the aggregate will enable the refractory to have a smaller pore structure than that of a sample made with coarser grains. In the case of fused cast

AZS, there are little to no pores in the refractory, which attribute to the good alkali resistance. In addition to the physical path the alkali follows when penetrating a refractory, a chemical reaction also occurs. This is where the mineralogy of the refractory sample is important. Some notable chemical reactions to consider are as follows:

1. $Na_2O + Al_2O_3 \cdot 2SiO_2 \rightarrow Na_2O \cdot Al_2O3 \cdot SiO_2$
   a. Nepheline; expansive – unstable above 2278°F (1248°C)
2. $Na_2O+Al_2O_3 \cdot SiO_2 \rightarrow Na_2O \cdot Al_2O_3 \cdot 6SiO_2$
   a. Albite; low melting point = 2019°F (1104°C)
3. $Na_2O + Al_2O_3 \rightarrow Na_2O \cdot Al_2O_3$
   a. Beta-alumina; expansive
   b. This is the most significant problem in superstructure applications. Formation results with expansion related problems: spalling, peeling, cracking as seen in alkali vapor tests.
4. $K_2O + Al_2O_3 \cdot SiO_2 \rightarrow K_2O \cdot Al_2O_3 \cdot 4SiO_2$
   a. Leucite; Expansive

Any of these expansive reactions can lead to spalling, peeling or cracking of the refractory lining. If albite, or any other low melting phase is formed, the refractory will begin to deteriorate as that phase is melted out of the refractory.

In the superstructure of a glass melting furnace, the hot-face refractory lining serves as structural elements for the furnace. Just like a house, if the walls begin to bow or the crown [roof] begins to sag, the life of the structure is shortened. Creep resistance is an important measure to determine if this bowing or sagging phenomenon is likely. The DIN creep test, 28.4 psi 50 hours at 1550°C was performed on all samples except sample Z; results are shown in Table 4. (Sample Z was too small to obtain a sample for creep testing in addition to other various tests performed in this study.)

Table 4. DIN creep test results, 28.4 psi 50 hours at 1550°C

	Sample A	Sample P	Sample V
0 to 50 hours	0.74	0.20	0.66
5 to 25 hours	0.31	0.09	0.30
20 to 50 hours	0.28	0.09	0.31

Due to limited availability of samples, only Sample A was able to be tested in thermal shock resistance as a comparison to fused cast AZS 34%. The high amount of glassy phase present in fused cast products, the bond structure is very brittle and causes the material to be highly susceptible to thermal shock. Generally speaking, this is not the case with sintered AZS products, as represented in the results of sample A in Table 5. Thermal shock resistance was tested using the prism spall method:

- For each sample, cycle three 2x2x3" (50.8x50.8x76.2 mm) prisms of refractory from 2200°F (1204°C) or 2650°F (1454°C) furnace to room temperature water and then air.
- The cycle consists of 10 minutes in the furnace, 2 minutes in water, 8 minutes in air, then back in the furnace.
- The furnace temperature is brought back up to the test temperature (takes a few minutes) and the cycle repeats.
- After each cycle, the replicates for each sample tested are evaluated quickly for extent of cracking.
- Once all three replicates reach failure, or 40 cycles are completed, the prism spall test is finished.
- The data is reported as an average of the three replicates for each sample.

Table 5. Prism Spall Results

	Sample A	Sample A	Fused Cast AZS 34%
Test Temperature	2200°F (1204°C)	2650°F (1454°C)	2200°F (1204°C)
Cycles to hairline crack	40+	20	1
Cycles to failure	40+	40+	2

Typically, the prism spall test is only performed at the lower temperature. However, in the case of Sample A, since it did not exhibit any cracking for all forty cycles, a higher test temperature was necessary to yield any cracking. Even at this higher test temperature, the sample did not fail, but it did exhibit some hairline cracking after twenty cycles. This is substantially better than the fused cast AZS 34% results, which exhibited complete failure after just two cycles at the lower test temperature.

Another test of great importance in material selection for superstructures is thermal conductivity. The thermal conductivity of a material is critical for designing the thickness of a refractory lining and dictates the level of insulation needed on a furnace. In the study, only sample A was tested since it will have such similar thermal conductivity values as the other sintered AZS samples. Figure 4 shows the comparison of sample A to fused cast AZS 34%.

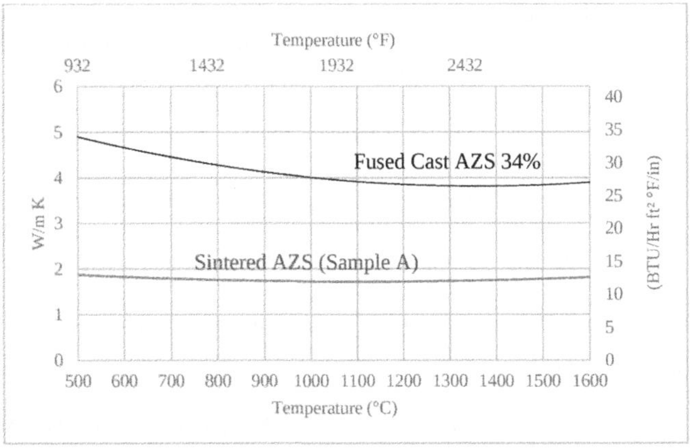

Figure 4. Thermal Conductivity of Fused Cast AZS 34% versus Sintered AZS

In the SEM/EDAX image of the fused cast AZS shown in Table 1, the continuous, interconnectivity of the grain boundaries and the very low amount of porosity attributes to the higher thermal conductivity values seen in Figure 4. The higher thermal conductivity, in combination with the fact that the shapes in the superstructure are typically produced with the gate scar intact, require furnace designers to use hot face lining thicknesses about 8-9" (203.2-228.6 mm). When building a superstructure with a hot face lining composed of sintered AZS, there is an opportunity to decrease the lining thickness and allow some flexibility in the insulation package offered. By reducing the hot face lining thickness, the change in heat flux is not the only benefit, but the cost of the refractory lining is also reduced.

ENHANCING PERFORMANCE BY INCREASING ENERGY EFFICIENCY

In the past section, discussion about the thermal conductivity difference between sintered AZS and fused cast AZS led to the conclusion that a design change in the lining thickness could be achieved. The question is how much does this effect the heat flux through the lining?

In order to gain a good understanding of the effect on cost and heat flux, four design models have been studied. All models were calculated using the same furnace parameters: 2800°F (1538°C) hot face temperature, 5 mph wind velocity, 70°F (21°C) ambient temperature and 0.5 cold face emissivity. The four models are outlined in Figures 5 through 8.

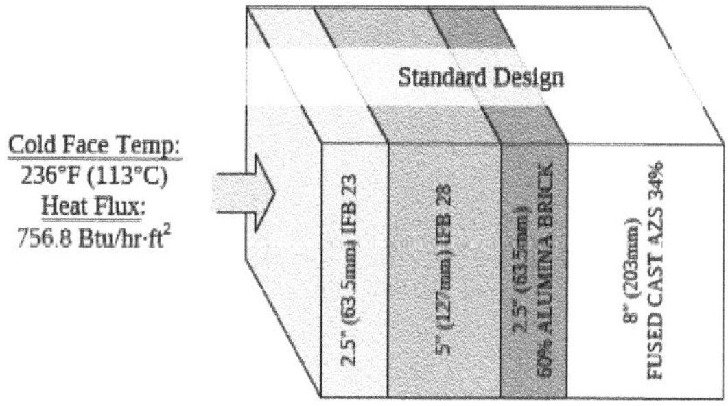

Figure 5. Model A; Standard superstructure design

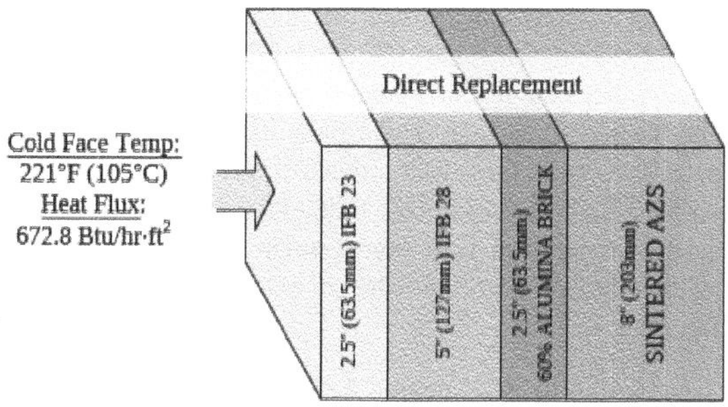

Figure 6. Model B; Direct replacement of sintered AZS in place of the fused cast AZS.

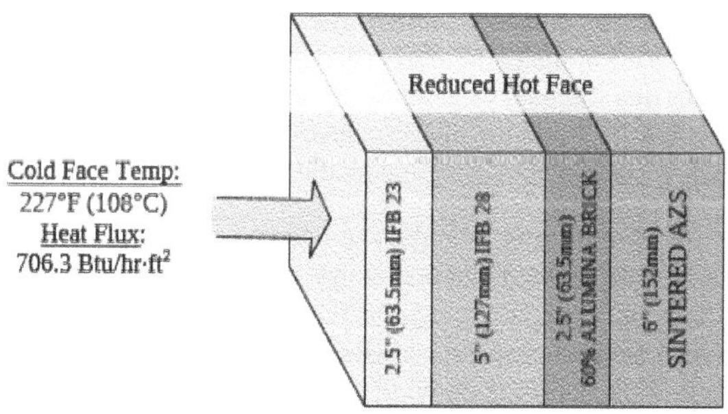

Figure 7. Model C; Reduced hot face thickness, same insulation package

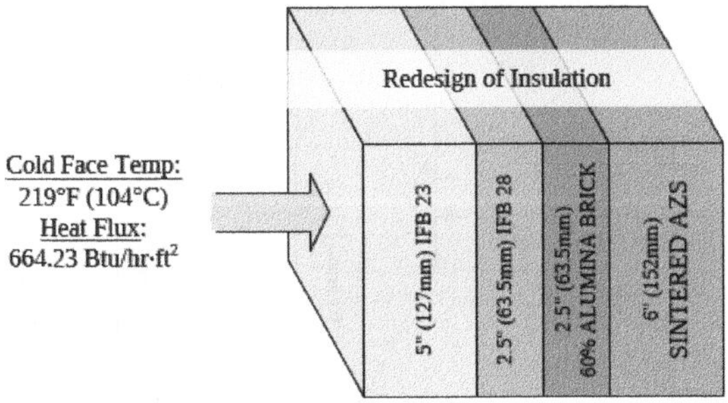

Figure 8. Model D; Redesign of insulation package

By only making a material swap as shown in Model B, the price for the lining will increase by 28%, but the energy savings will be 11%. Model C takes advantage of sintered AZS material being void free and reduces the hot face material thickness from eight inches to six inches, but maintains the same insulation package. This results in the same relative cost as Model A, but has an energy savings of 7%. Out of these four models, the most economic and energy efficient design is achieved when changing both the hot face lining thickness and the insulation package as shown in Model D: 1% cost savings and 12% energy savings. These four models are used as relative examples to show the effect on heat flow when of using a sintered AZS hot face material versus a fused cast AZS option.

Another benefit of utilizing a sintered AZS in a superstructure in lieu of the fused cast AZS is the ability to coat the hot face with a high emissivity coating. Results have shown that when a soda-lime container glass furnace has a high emissivity coating on just the crown, it typically has a 3-4% energy savings as compared to the furnace without the coating. However, furnaces that have had the ability to coat the entire crown and superstructure have demonstrated 6-8% energy savings. This means that if model D, as shown in image 6, were to have a high emissivity coating applied to the sintered AZS and the other superstructure materials, then energy savings could be as high as 20% when compared to using the model A design in the same furnace.

A high emissivity coating in a glass furnace must have the ability to reach temperatures greater than 2600°F (1427°C) without degrading or losing its emissivity value. The most utilized high emissivity coating used in this application is Emisshield®. This product was originally developed by NASA to protect space vehicles. In these applications, the coating is designed to modify the surface of the substrate on which it is placed to increase the emissivity and therefore the amount of energy radiated from it. In space applications, temperatures can range from subzero to over 3000°F (1649°C) in a matter of seconds. The coating was designed to maintain adhesion to a space vehicle's hull without losing its emissivity under these conditions. This coating is applied just 2-3 mils thick – the thickness of a garbage bag – and is capable of operating at temperatures up to 3500°F (1927°C).

This coating has been used in over 100 furnaces world-wide and the results of a few of these furnaces was monitored very closely, collecting a lot of data, which was published in the 71st and 74th Conference on Glass Problems. There are several benefits of using a high emissivity coating that glass makers have taken advantage of more even heating, to longer refractory life, to increased productivity. But in glass furnaces, the main drivers have been increased fuel savings, decreased NOx and reduced carbon footprint. These latter benefits demonstrate why the use of a high emissivity coating is a critical element to having a high performance superstructure.

CONCLUSION

Each furnace design has its own considerations to account for; so working closely with an experienced refractory design team in advance of a furnace build can help determine the best method for achieving energy savings, maintaining or reducing the cost of a lining, and of course ensuring good glass quality and long campaign life. Historically superstructures in container glass furnaces have been designed using fused cast AZS materials. Newer products on the market enable furnace engineers to broaden the material selections, which allow them to incorporate larger shapes and utilize the latest technology in emissivity coatings. By balancing refractory lining cost, glass quality, energy savings, and furnace campaign life, furnace engineers are capable of designing the next level of high performance superstructures.

REFERENCES

1. Kleeb, T. and Fausey, B. (2011) Fuel Savings with High Emissivity Coatings, in 71st Conference on Glass Problems: Ceramic Engineering and Science Proceedings, Volume 32 (ed C. H. Drummond), John Wiley & Sons, Inc., Hoboken, NJ, USA. doi: 10.1002/9781118095348.ch13
2. Kleeb, T. and Fausey, B. (2014) High Emissivity Coatings in Glass Furnaces, in 74th Conference on Glass Problems (ed S. K. Sundaram), John Wiley & Sons, Inc., Hoboken, NJ, USA. doi: 10.1002/9781118932964.ch23
3. Ratto, P. C. (2013) The end of fused-cast refractories in soda-lime glass furnace superstructures? *Glass International, February.* pp. 29-31.
4. van Dijk, G. & Arnoldus, F. Glass defects originating from glass melt/fused cast AZS refractory interaction. Diss. Thesis, 1994.

ENHANCED HYDRATED LIME – A SIMPLE SOLUTION FOR ACID GAS COMPLIANCE

G. Hunt
Lhoist North America
Fort Myers, FL, USA
M. Sewell
Lhoist North America
Fort Worth, TX, USA

ABSTRACT

The landscape of environmental regulations in the US is complex and continues to evolve. State and Federal regulations as well as air permits mandate increasingly more stringent emission requirements for various air pollutants for both new and existing facilities for a number of acid gas species (i.e. SO_2, HCl, HF, SO_3) as well as Hg and particulate matter. As a result of the need to comply with these stringent emission limits, there is a growing desire for a low cost/easy-to-install solution. DSI technology can offer that as it is a low capital cost solution with a relatively small equipment footprint, low power consumption and easy to retrofit to a majority of existing facilities. As DSI technology has matured the systems have become more reliable and advancements in calcium based sorbents have provided new compliance solutions that weren't available in years past. Optimization of the physical properties of Lhoist's Generation 2 and 3 Enhanced Hydrated Lime products have demonstrated upwards of 90% SO_2 and 95+% HCl / HF reduction with DSI technology over a range of applications. This paper will present data from DSI demonstration testing performed for SO_2 and HCl control at a variety of applications using Lhoist's enhanced hydrated lime products.

INTRODUCTION

Environmental regulations for the industrial and utility sectors present challenges for these facilities to find an economical compliance solution. Implementation of the MATS and IB MACT rules as well as state specific regulations and consent decrees are driving the requirement for facilities to find a multi-pollutant emission control solution. These regulations and agreements will require increased emission reductions for a number of acid gas species (i.e. SO_2, HCl, HF, SO_3), Hg and particulate matter. DSI and ACI in combination with BH filters or ceramic filters have been viewed as the choice technologies at several facilities to comply with these regulations as they offer a low capital cost solution with a relatively small equipment footprint, low power consumption and the ability to easily retrofit a majority of existing facilities compared to alternative technologies.

DESCRIPTION OF DSI TECHNOLOGY

DSI is a mature technology that has seen significant growth since the early 2000s by utilities in the USA requiring SO_3/H_2SO_4 emissions reduction for mitigation of a visible blue stack plume. DSI consists of a storage vessel (typically a silo), dosing system, motive air system, conveying piping, sorbent portioning devices and injection lances. An alkaline sorbent is unloaded into the storage vessel where it flows into the dosing system. The dosing system typically consists of weigh hoppers, screw feeders and/or rotary valves which meter the sorbent into the piping system in a controlled manner. The motive air system typically consists of a blower and after-cooler to pressurize ambient air and subsequently cool the conveying air which is then used to transport the sorbent from the dosing system discharge through the conveying piping to the partitioning device(s). The partitioning device(s) are typically consists of a tee section and/or flow splitter to disperse the sorbent to each injection lance. The injection lances are commonly a straight pipe cut to a desired length to penetrate into the flue gas stream so that the sorbent is strategically placed into contact with the flue gas stream. Figure 1 is a generic schematic of a typical DSI system.

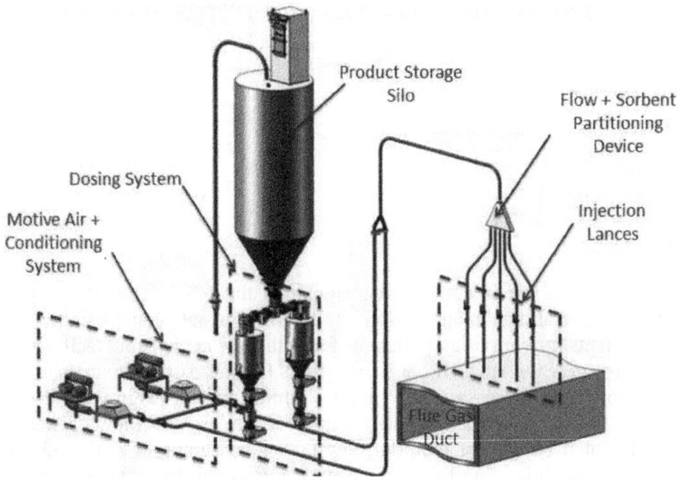

Figure 1. Generic Schematic of Typical DSI System

DSI offers the following benefits over other acid gas control technologies (i.e. WFGD, SDA, CDS):

- Lower installed capital cost
- Relatively easy to retrofit to a majority of facilities (only injection lances are in direct contact with exhaust gas)
- System has good process flexibility for various sorbents and ability to easily modulate based on unit load and/or different fuels
- Small equipment footprint (typically footprint of one or two silos and blower building)
- Relatively short design, fabrication and installation schedule as there is approximately one year schedule from contract award to commercial operation
- Low consumable requirements (i.e., air and water) as well as low parasitic power requirements
- Capability to easily perform full scale trial to validate technology capabilities and determine design basis for system design

Over the past few years there have been significant design improvements to DSI systems based on operating experiences from past installations, which have increased current DSI system reliability and availability. Additionally, improved sorbent injection technologies have been developed and are being implemented to reduce sorbent consumption up to 40%[1]. CFD modeling capabilities have advanced to include not only sorbent dispersion modeling but also reaction kinetics and DSI performance predictive capabilities[2, 3, 4]. Overall, experience with DSI has significantly increased and collection and analysis of DSI performance data have helped provide better understanding on the impacts the process conditions (i.e. gas temperature, moisture, etc.) play in DSI chemistry.

Concurrently, there have also been developments to improve the performance of DSI sorbents (i.e., enhanced hydrated lime sorbents) such that these improved sorbents provide a given level of acid gas removal at lower sorbent injection rates or alternatively an improved performance previously unattainable can now be reached. This paper discusses the development and application of Generation 2 and 3 Enhanced Hydrated Lime products used in conjunction with DSI technology to provide effective acid gas removal.

ENHANCED HYDRATED LIME DEVELOPMENT

The first generation of enhanced hydrated lime sorbents (designated by Lhoist as Sorbacal® A) was developed in the 1980s by increasing the surface area of hydrated lime significantly above 10-20 m^2/g as is normal in standard hydrated lime. The high surface area combined with a small particle size gave Generation 1 Enhanced Hydrated Lime a significant performance boost compared to standard hydrated lime. During the acid removal reaction, the rate is slowed down because the reaction products, such as $CaSO_4$, form a diffusion layer on the fresh unreacted $Ca(OH)_2$ material. More important, the reaction product $CaSO_4$ has a higher molar volume and thus gradually fills up the porosity of the sorbent. This means that to improve the performance, increasing the surface area alone is not sufficient; the pore volume also needs to be increased.

Extensive research by the Lhoist group in the 1990s showed that indeed both the capture capacity and the reactivity of the sorbent are directly proportional to the pore volume (refer to Figure 2). In contrast, the surface area was found to be contributing to a lesser extent to the acid gas removal efficiency. This research led to the development of a second generation of sorbents, designated as Sorbacal® SP, with both a higher pore volume (> 0.2 cm^3/g) and a higher surface area (> 40 m^2/g), which are two to three times higher than that of a standard hydrated lime. Laboratory scale, pilot scale and commercial scale tests have demonstrated that the reactivity of Generation 2 Enhanced Hydrated Lime may be twice or more than that of high quality hydrated lime.

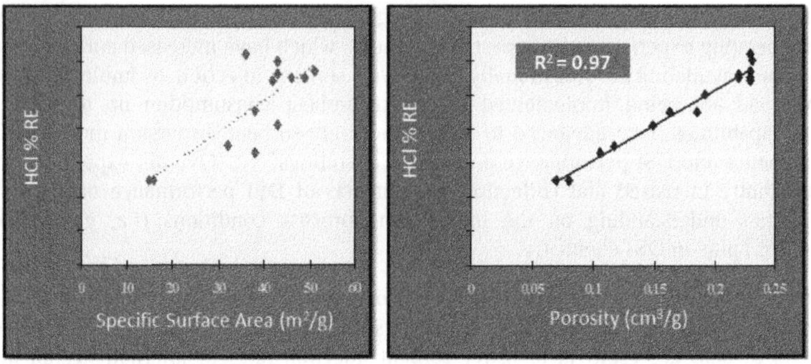

Figure 2. Pilot Testing Results Correlating Removal Efficiency with Hydrated Lime Specific Surface Area and Porosity

The third generation of sorbents, designated as Sorbacal® SPS, combines the engineered pore structure properties of Generation 2 Enhanced Hydrated Lime with a chemical modifier that provides an additional acceleration of the reaction rate. Additionally, the chemical modifier used for Generation 3 Enhanced Hydrated Lime production has demonstrated improved response for ESPs compared to other hydrated lime products, due to its lower resistivity[5]. Figure 3 illustrates pilot SO_2 testing work conducted to illustrating Generation 3 Enhanced Hydrated Lime increased reactivity over Generation 2 Enhanced Hydrated Lime despite comparable pore volume. Today, Lhoist operates six Enhanced Hydrated Lime manufacturing locations in Europe, has licensed the technology to five Japanese plants and has three manufacturing locations in the USA. Figure 4 shows the most pertinent physical and chemical properties of the various commercially available hydrated lime products discussed in this section of the paper.

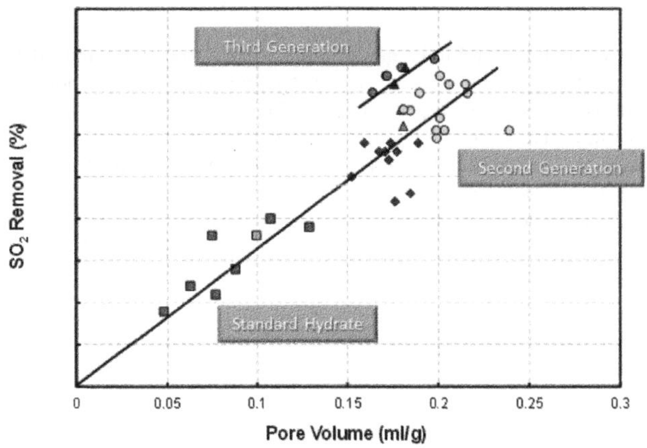

Figure 3. Pilot SO_2 Test Results Illustrating Relative Sorbent Reactivities

			Generation 2	Generation 3
Sorbent	Standard Hydrated Limes	FGT Grade Sorbacal® H	Sorbacal® SP	Sorbacal® SPS
Figure		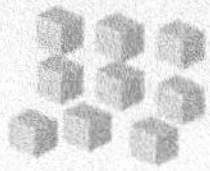	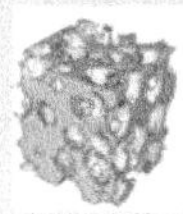	ACTIVATION 
Typical Available $Ca(OH)_2$ - [%]	92 – 95	93	93	93
Typical Surface Area - [m^2/g]	14 – 18	20	40	40
Typical Pore Volume - [cm^3/g]	~0.07	0.08	0.20	0.20
Typical D_{50} - [microns]	5 – 7	5 – 7	8 – 12	8 – 12

Figure 4. Properties of Various Hydrated Lime Products Available

ENHANCED HYDRATED LIME BENEFITS

The Generation 2 and 3 Enhanced Hydrated Lime product properties previously described make it more reactive than standard hydrated lime; meaning equal acid gas removal can be achieved at lower sorbent injection rates. Lower sorbent injection rates may equate to an operating cost savings associated with the annual spend on sorbent. However, sorbent economics are a function of not only sorbent material pricing but also the freight cost as well as the relative sorbent consumption which all must be considered when evaluating which sorbent provides the most economical solution from an operating cost perspective. One metric recommended to evaluate the relative sorbent cost as a function of sorbent delivered price and consumption is calculating the cost per mass of acid gas removed as it considers both cost and relative sorbent reactivity. Figure 5 is an economic case study from an actual sorbent bid process for an industrial application comparing Generation 2 Enhanced Hydrated Lime and a standard hydrated lime which were approximately equal distance from their relative manufacturing locations to the customer site. Figure 5 shows how Generation 2 Enhanced Hydrated Lime was more cost effective for this site when the cost and relative reactivity of each sorbent is taken into consideration.

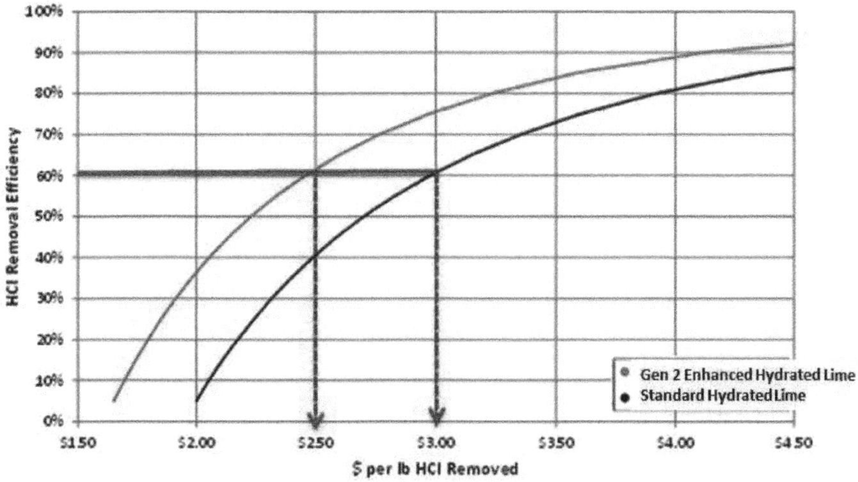

Figure 5. Case Study Comparing Generation 2 Enhanced Hydrated Lime and Standard Hydrated Lime Sorbent Economics

Additional benefits when utilizing Enhanced Hydrated Lime products in lieu of standard hydrated lime include the following:

- *Storage silo capacity* – If a system is designed based on standard hydrated lime and later converted to an Enhanced Hydrated Lime product, the reduced injection rate will translate to increased storage silo capacity in terms of number of storage days. However, if a system is designed based on an Enhanced Hydrated Lime product then this can result in a capital cost savings in the DSI system design by utilizing a smaller storage silo.

- *Reduced impacts on ESP/BH filter and waste handling equipment* – Reduced sorbent injection rates will result in less balance of plant impacts. Less sorbent injected into an ESP will have fewer impacts on dust resistivity entering the ESP. Less sorbent injected in a BH filter will result in slower accumulation of filter cake reducing the frequency of filter bag cleaning.

- *Less residue generated / lower disposal costs* – Injecting at lower sorbent dosing rates will result in lower generation of residue / byproducts. If the residue is disposed of then this will result to reduced disposal costs.

- *Fewer deliveries* – Reduced sorbent consumption will result in fewer deliveries. This will make logistics and planning easier and reduce the potential for late deliveries and/or demurrage charges.

- *Achieve higher performance* – In addition to reduced sorbent consumption there may be a removal efficiency achieved that is not achievable with standard hydrated lime. This may be important as regulations and/or air permits become more stringent.

FULL SCALE OPERATING DATA

Lhoist has been actively participating in full scale demonstrations with Enhanced Hydrated Lime products in the US over the past four years and for decades in Europe (specifically on glass applications) to demonstrate the acid gas removal efficacy capabilities and improved relative performance versus other commercially available sorbents. Enhanced Hydrated Lime products have been utilized in not only the EGU industry but also several industrial sectors such as glass, cement[6, 7, 8], pulp and paper, tile, waste incineration[9] to name a few. This section of the paper will present data from multiple full scale DSI trials to present the capabilities of DSI technology with Enhanced Hydrated Lime products to provide multi-pollutant control that could be further utilized in the glass industry.

SO_2 is typically the most common acid gas pollutant requiring reduction in the glass industry. The first case study presents data from a full scale trial at an EGU burning a low sulfur coal equipped with an ESP to capture filterable particulate emissions. This EGU tested a variety of sorbents including Generation 3 Enhanced Hydrated Lime and a standard hydrated lime. Both sorbents were injected at the same injection location (flue gas temperature was approximately 700°F) using the same quantity and configuration of injection lances. The results were normalized by calculating the NSR, which is a ratio of moles of calcium injected to moles of uncontrolled SO_2, at each condition in order to provide a consistent comparison between the sorbents. Figure 6 provides a graphical summary of the results which illustrate that Generation 3 Enhanced Hydrated Lime was approximately 40% more effective than standard hydrated lime at this facility. Figure 6 also shows that DSI with Generation 3 Enhanced Hydrated Lime was capable of achieving high levels of SO_2 control, approximately 85% SO_2 removal achieved. Generation 3 Enhanced Hydrated Lime had the capability to achieve higher SO_2 removals (which was demonstrated at lower boiler load conditions); however, higher injection rates were not evaluated since 85% SO_2 removal exceeded the facility's target SO_2 removal efficiency.

DSI technology is also commonly used to control HCl emissions. LNA participated in a full scale demonstration to prove that DSI with Generation 2 Enhanced Hydrated Lime was capable of achieving over 99% HCl removal from a medical waste incinerator where inlet HCl emissions were extremely variable and compliance was required over a tight averaging period (3 hour rolling average). This facility had a PJFF for particulate control and Generation 2 Enhanced Hydrated Lime was injected into the exhaust gas stream at a temperature of 500-600°F. The facility had continuous HCl monitors upstream of the sorbent injection ports as well as the stack so real-time HCl removal efficiency could be calculated. Figure 7 shows the real time results over the course of a week of continuous sorbent injection. In Figure 7 the red lines show the inlet HCl concentrations and illustrate the highly variable nature of the medical waste incinerated as HCl emissions from the incinerator ranged from approximately 500 to 5,000 ppmv (dry) corrected to 7% O_2 and the green line shows the HCl removal efficiency required to achieve the 6.6 ppmv (dry) corrected to 7% O_2 limit. The blue line shows the actual HCl removal achieved by injecting Generation 2 Enhanced Hydrated Lime with the DSI system.

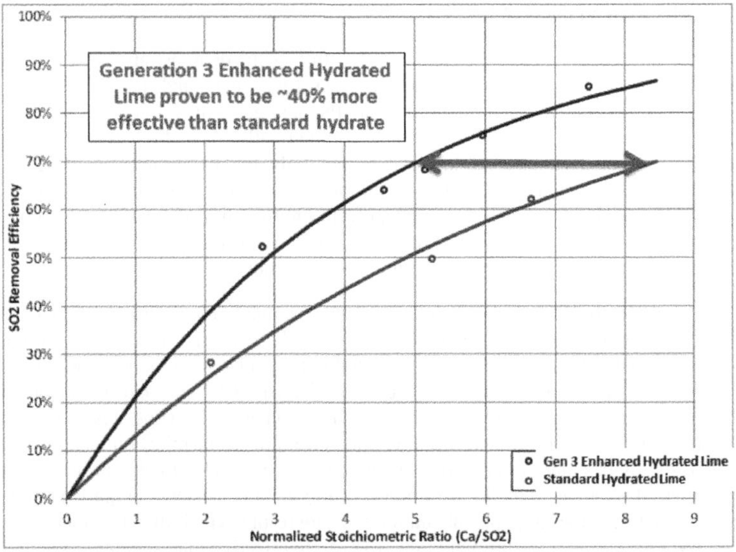

Figure 6. SO_2 Removal vs. NSR for Standard Hydrated Lime and Generation 3 Enhanced Hydrated Lime product at EGU

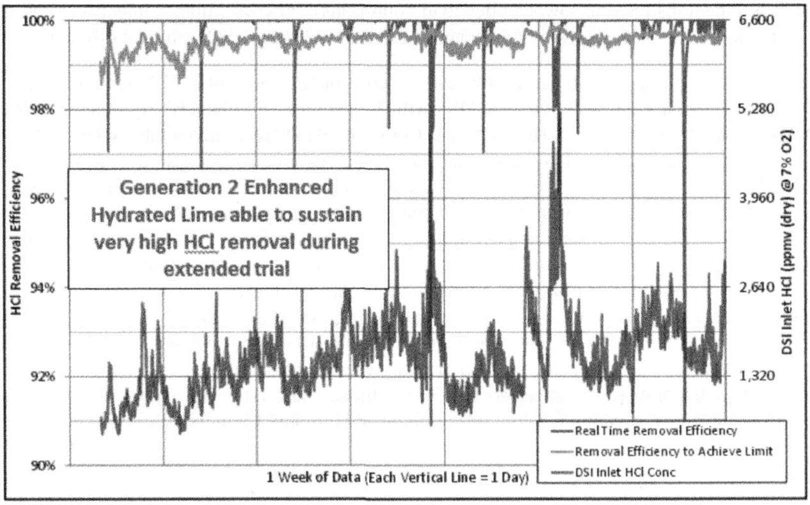

Figure 7. Real-Time HCl Results from Week-Long DSI Trial with Generation 2 Enhanced Hydrated Lime

Additionally, DSI using Enhanced Hydrated Lime products have also provided effective high level HF removal efficiency. A tile manufacturing facility performed a week-long trial for multi-pollutant control to demonstrate SO_2, HCl and HF abatement using Generation 3 Enhanced Hydrated Lime. HF emissions at the stack and upstream of the injection location were continuous monitored using FTIR monitors. Generation 3 Enhanced Hydrated Lime was injected into the exhaust gas using a single injection port where gas temperature was approximately 300-400°F. Figure 8 shows the real-time inlet and stack HF emissions throughout the trial. Despite modulation of the Generation 3 Enhanced Hydrated Lime injection rates in accordance with various testing activities, the HF emissions at the stack remained at very low concentrations, approximately 98+% HF removal.

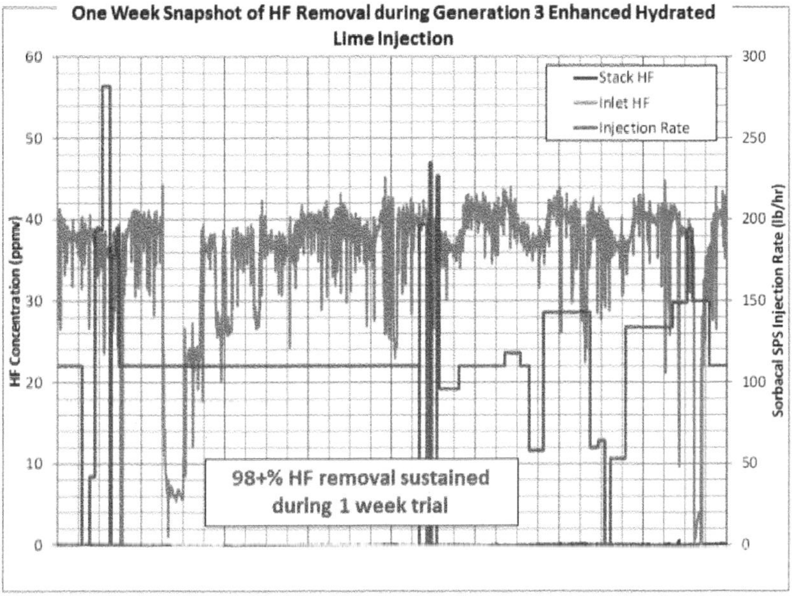

Figure 8. Real-Time HF Emissions Monitoring during Generation 3 Enhanced Hydrated Lime Injection at Tile Facility

Another benefit of using DSI technology with Enhanced Hydrated Lime Products is the ability to achieve high selenium removal efficiency. Since some glass plants recycle residue back into their raw materials and some glass products utilize selenium for production, Enhanced Hydrated Lime products can help to reduce costs necessary to introduce selenium to the batch. Figure 9 is selenium removal data from a float glass facility in Europe which compares Generation 3 Enhanced Hydrated Lime selenium removal against sodium bicarbonate, another alkaline sorbent used in DSI. Figure 9 shows that Generation 3 Enhanced Hydrated Lime is more reactive than sodium bicarbonate for selenium removal and was able to achieve 95% selenium removal.

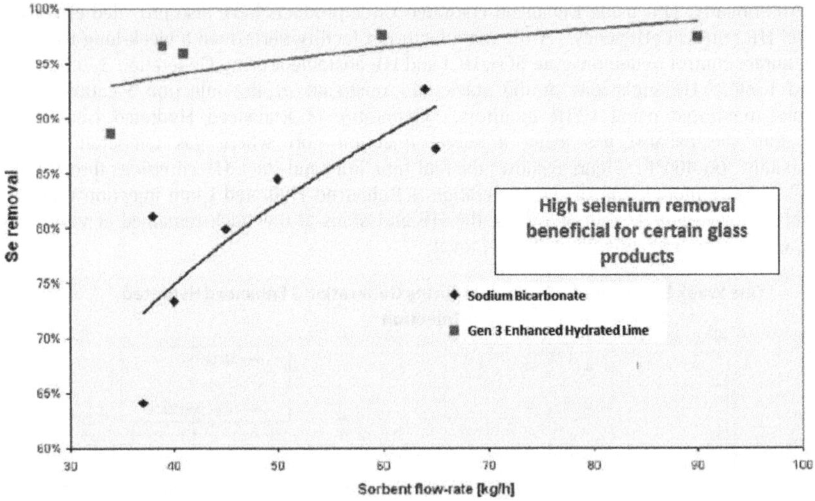

Figure 9. Relative Selenium Removal Efficiency of Generation 3 Enhanced Hydrated Lime and Sodium Bicarbonate from Float Glass Plant

KEY TAKE AWAYS

The purpose of this paper is to present information that shows that DSI technology coupled with Lhoist's Enhanced Hydrated Lime products provides the glass industry a simplistic solution to control acid gas emissions. DSI technology is a mature, low capital expenditure system that is flexible to process conditions and easy to retrofit to a majority of facilities.

While advancements have been made in DSI systems to ensure they are highly reliable pieces of equipment, advancements have also been made in sorbents. Generation 2 and 3 Enhanced Hydrated Lime products are produced with higher surface area and pore volume than standard hydrated lime products which has provides increased reactivity towards acid gases. Generation 2 and 3 Enhanced Hydrated Lime products have demonstrated high SO_2, HCl, HF and SO_3 removal efficiencies at EGU and a multitude of industrial applications. Using Generation 2 and 3 Enhanced Hydrated Lime have provided other benefits that include reduced sorbent operating costs, increased storage silo capacity, fewer deliveries as well as lower disposal costs, less balance of plant impacts on ESPs, BH filters and residue handling systems.

DSI with Generation 2 and 3 Enhanced Hydrated Lime provides a cost effective solution for the glass industry to reduce acid gas emissions in order to comply with regulatory and air permit requirements. Currently, Generation 2 Enhanced Hydrated Lime is utilized at about a dozen US glass facilities to successfully help control acid gas emissions and can be a cost efficient solution as regulations continue to become more stringent in this industry.

ABBREVIATIONS

ACI	Activated carbon injection
BH	Baghouse filter
CDS	Circulating dry scrubber
DSI	Dry sorbent injection
EGU	Electric generating utility
ESP	Electrostatic precipitator
FGT	Flue gas treatment
FTIR	Fourier transform infrared
Hg	Mercury
IB MACT	Industrial Boiler Maximum Available Control Technology
LNA	Lhoist North America
MATS	Mercury Air Toxics Standards
NSR	Normalized stoichiometric ratio
PJFF	Pulse jet fabric filter
SDA	Spray dryer absorber
WFGD	Wet flue gas desulfurization

REFERENCES

1. Nick Evans, Mitch Lund. *Using Sorb-Tec™ Technology to Reduce Costs and Improve Performance for DSI and ACI Systems.* Council of Industrial Boiler Operator Industrial Emissions Control Technology XIV, August 2016.
2. Rodney Foo, Rob Berger, Johan Heiszwolf. *Reaction Kinetic Modeling of DSI for MATS Compliance.* MEGA Symposium, August 2016.
3. Guisu Liu, Jose Sanchez, Greg Archer. *Optimizing Dry Sorbent Injection Performance Using Chemistry-Based CFD Modeling.* MEGA Symposium, August 2016.
4. Johan Heiszwolf, Olivier Nyssen. *Improving Dry Sorbent Injection with Computational Fluid Dynamics.* NPT Procestechologie, pages 12-13, February 2016.
5. Rodney Foo, Jim Dickerman, Jerry Hunt, Landon Johnson, Johan Heiszwolf. *ESP Compatible Calcium Sorbent for SO_2 Capture at Great River Energy's Stanton Station.* MEGA Symposium, August 2016.
6. Melissa Sewell, William Ciro. *HCl Control for MACT Compliance.* World Cement Magazine, April 2014.
7. Melissa Sewell, Jerry Hunt. *Optimising Dry Sorbent Injection Technology.* World Cement Magazine, April 2015.
8. Gerald Hunt. *The Solution to SO_2 Pollution.* International Cement Review, July 2016.
9. Gerald Hunt, Melissa Sewell. *Utilizing Dry Sorbent Injection Technology to Improve Acid Gas Control.* 34th International Conference on Thermal Treatment Technologies & Hazardous Waste Combustors, October 2015.

Printed and bound by CPI Group (UK) Ltd, Croydon, CR0 4YY

17/09/2026

14964241-0001